IoT-Based Smart Waste Management for Environmental Sustainability

Smart and Intelligent Computing in Engineering

Series Editor:
Prasenjit Chatterjee, Morteza Yazdani, Dragan Pamucar, and Dilbagh Panchal

Artificial Intelligence Applications in a Pandemic
COVID-19
Salah-ddine Krit, Vrijendra Singh, Mohamed Elhoseny, Yashbir Singh

Advanced AI Techniques and Applications in Bioinformatics
Loveleen Gaur, Arun Solanki, Samuel Fosso Wamba, Noor Zaman Jhanjhi

IoT - Based Smart Waste Management for Environmental Sustainability
Biswaranjan Acharya, Satarupa Dey, and Mohammed Zidan

Applications of Computational Intelligence in Concrete Technology
Sakshi Gupta, Parveen Sihag, Mohindra Singh, and Utku Kose

For more information about this series, please visit: https://www.routledge.com/our-products/book-series

IoT-Based Smart Waste Management for Environmental Sustainability

Edited by

Biswaranjan Acharya, Satarupa Dey, and Mohammed Zidan

CRC Press
Taylor & Francis Group
Boca Raton London New York

CRC Press is an imprint of the
Taylor & Francis Group, an **informa** business

First edition published 2022
by CRC Press
6000 Broken Sound Parkway NW, Suite 300, Boca Raton, FL 33487-2742

and by CRC Press
4 Park Square, Milton Park, Abingdon, Oxon, OX14 4RN

CRC Press is an imprint of Taylor & Francis Group, LLC

© 2022 Taylor & Francis Group, LLC

Library of Congress Cataloging-in-Publication Data

Names: Acharya, Biswa Ranjan, editor. | Dey, Satarupa, editor. | Zidan, Mohammad, editor.
Title: IoT-based smart waste management for environmental sustainability / edited by Biswaranjan Acharya, Satarupa Dey, and Mohammed Zidan.
Description: First edition. | Boca Raton, FL: CRC Press, 2022. | Series: Smart and intelligent computing in engineering | Includes bibliographical references and index.
Identifiers: LCCN 2021056623 (print) | LCCN 2021056624 (ebook) | ISBN 9781032013916 (hbk) | ISBN 9781032025971 (pbk) | ISBN 9781003184096 (ebk)
Subjects: LCSH: Integrated solid waste management--Automation. | Refuse and refuse disposal--Automation. | Internet of things.
Classification: LCC TD794.2.I67 2022 (print) | LCC TD794.2 (ebook) | DDC 363.72/85--dc23/eng/20220119
LC record available at https://lccn.loc.gov/2021056623
LC ebook record available at https://lccn.loc.gov/2021056624

ISBN: 978-1-032-01391-6 (hbk)
ISBN: 978-1-032-02597-1 (pbk)
ISBN: 978-1-003-18409-6 (ebk)

DOI: 10.1201/9781003184096

Typeset in Palatino
by Deanta Global Publishing Services, Chennai, India

Contents

Preface

Environmental sustainability is a vital topic that emphasizes the protection of environmental reserves/resources and their conservation for future generations. Environmental sustainability is a sensitive matter that guides all aspects of life. The development of a community with environmental consciousness for the management of resources, renewable energy, and protection of air and water quality highlights the concept of sustainable development. The focus should be made on development to meet the needs of people without compromising or harshening life for future generations. However, it is challenging to maintain the environment with growing urbanization. Therefore, the practice of the concept of sustainable development is the way forward for a fairer world without damage to the environment. In relation to all aspects of our lives, it needs ideas and initiatives to promote certain vital projects for a sustainable environment, including the recycling of valuable resources.

Smart systems and technologies like the Internet of Things (IoT) are concepts in which surrounding objects are connected through wired and wireless networks without user intervention. In the field of IoT, the objects communicate and exchange information to provide advanced intelligent services for users. Owing to the recent advances in mobile devices equipped with various sensors and communication modules, together with communication network technologies such as Wi-Fi and LTE, the IoT has created considerable academic interest.

Sustainable management and mitigation of waste products have remained a challenge to most industries and society. Waste management service providers and cities can benefit from IoT-powered smart waste management solutions. Using this technology, waste management agencies can increase their operational efficiency, cut costs, and enhance customer satisfaction. With the decrease in the implementation cost of wireless technologies, there are plenty of technologies available that make smart waste management solutions possible.

This book aims to consolidate and summarize the cutting-edge smart technologies that deal with different aspects of generation, management, mitigation, and recycling of waste products for a sustainable environment using the latest computational technologies like IoT, edge, and artificial intelligence. It will enhance the knowledge of these systems for academics and/or the professional community working in the interfaces of environmental science and computer science.

This book consists of ten chapters contributed by different groups of researchers working on different aspects of IoT and sustainable waste management. Chapters 1 and 2 deal with different aspects of waste generation and management of e-waste. As there is no concrete policy or initiatives implemented for the management of e-waste, it is mainly dumped in an unorganized way. Use of IoT in e-waste management can provide an appropriate alternative to the management of e-waste. In Chapter 3, other aspects such as the Smart Waste Bin Management system, its monitoring, and tracking the collection of waste along with timely transportation to recycling units and maintenance of the storage spaceare discussed in detail. Chapter 4 mainly deals with the management of food and agricultural waste and the use of artificial intelligence systems which assist in reducing waste during harvest and postharvest processes. Chapter 5 mainly deals with the municipal solid waste (MSW) which is a vital aspect of sustainable waste management in different cities of India as the case study. IoT applications such as radio-frequency identification (RFID), cameras, and cloud computing (data assimilation and communication) in MSW management have been used for effective collection, segregation, and transportation and provide better solutions for recycling/composting and contributing toward economic growth. Waste generation from the health care sector is also something which needs proper management and control, as it generates a significant amount of infectious and/or toxic waste, classified as hazardous waste. Chapter 6 represents another IoT-based innovative smart dustbin which automatically collects waste from appropriate places and also disposes to appropriate places using ultrasonic sensor, GMS and GPS module. Chapter 7 describes conceptual of wastes and types with effective solid waste management in general. It also uses different sensors and RFID for effective management. Chapter 8 represents serverless IoT architecture for smart waste management through cloud computing. Chapter 9 mainly deals with hospital solid waste generation and management. IoT plays a vital role in appropriately handling this waste, with the use of connected sensor devices and a planned IoT-based environment. It helps in identification based on images and gathering data on their own, and also removes possible human errors. The last chapter presents a deep learning-based garbage classification system that can be run in the IoT environment. Chapter 10 focuses auto classification of garbage using deep learning technique.

Finally, I would like to acknowledge the efforts of all the contributors for their continuous effort in bringing the book to fruition. The continued assistance of the Editorial Department of CRC Press is also highly appreciated.

The MathWorks, Inc.

3 Apple Hill Drive

Natick, MA 01760-2098 USA

Tel: 508 647 7000

Fax: 508-647-7001

E-mail: info@mathworks.com

Web: www.mathworks.com

Editors

Biswaranjan Acharya is an academic currently associated with Kalinga Institute of Industrial Technology Deemed to be University, Bhubaneswar, India, and is pursuing a doctorate in Computer Application at Veer Surendra Sai University of Technology (VSSUT), Burla, India. He has earnedan MCA from IGNOU, New Delhi, India, and M.Tech in Computer Science and Engineering from Biju Patnaik University of Technology (BPUT), Rourkela, India. He is also associated with various educational and research societies such as IEEE, IACSIT, CSI, IAENG, and ISC. He has worked in industry as a software engineer and in academia at Ravenshaw University, Cuttack, India. He is currently doing research in fields such as data analytics, computer vision, machine learning, and IoT. He owns eight patents and has published several research articles in internationally reputed journals.

Satarupa Dey is assistant professor in the Department of Botany, University of Calcutta, Kolkata, India. She completed her doctorate in 2012 in bioremediation of toxic metals. She was a research fellow in the Department of Biotechnology, Government of India, and later worked as a visiting scientist at the Agricultural and Ecological Research Unit of the Indian Statistical Institute, Kolkata, India. She is the recipient of the Woman Scientist (WOSA) fellowship of the Department of Science and Technology, Government of India. She has published in several journals of national and international repute.

Mohammed Zidan earned his Ph.D. degree from Sohag University, Egypt, in 2019. He works as a postdoctoral fellow at the Center for Photonics and Smart Materials (CPSM), Zewail City of Science and Technology, Giza, Egypt. He is a research consultant at Engineering College, Abu Dhabi University, UAE. He has published several papers and has two patents in machine learning and IoT. His current research interests include machine learning, IoT, quantum machine learning, quantum computing, and quantum IoT.

Contributors

Ibrahim A. Abouelsaad
Faculty of Desert Agriculture
King Salman International
University
South Sinai, Egypt

Gazala Yasmin Ashraf
Amity University
Raipur, India

Nivedita Das
KIIT University
Bhubaneshwar, India

Emad H. El-Bilawy
Faculty of Desert Agriculture
King Salman International
University
South Sinai, Egypt

Islam El-Sharkawy
Florida A&M University (FAMU)
Center for Viticulture & Small Fruit
Research
United States

Hrishikesh Chandra Gautam
Department of Civil Engineering
Indian Institute of Technology
Madras
Chennai, India

Earnest Paul Ijjina
Department of CSE
NIT
Warangal, India

Rampavan Medipelly
NIT
Warangal, India

Alok Narayan
KIIT Deemed to be University
Bhubaneswar, India

Sankalp Nayak
KIIT Deemed to be University
Bhubaneswar, India

Nitika Rani
Institute of Environmental Studies
Kurukshetra University
Kurukshetra, India

Babita Panda
Kalinga Institute of Industrial
Technology
Bhubaneswar, India

Gyanendra Kumar Panda
Tata Consultancy Services
Bhubaneshwar, India

Jyotiprakash Panigrahi
KIIT University hubaneshwar,
India

Arjyadhara Pradhan
School of Electrical Engineering
Kalinga Institute of Industrial
Technology
Bhubaneshwar, India

Uma Rahangdale
Amity University
Raipur, India

Jitendra Kumar Rout
NIT
Raipur, India

Bholanath Roy
Computer Science and Engineering
Department
Maulana Azad National Institute of
Technology
Bhopal, India

Chandrima Roy
KIIT University
Bhubaneshwar, India

Sarita Samal
School of Electrical Engineering
Kalinga Institute of Industrial
Technology
Bhubaneshwar, India

Amar Shinde
Department of Civil Engineering
Manipal Institute of Technology
Manipal Academy of Higher
Education
Manipal, India

Rahul Singh
Computer Science and Engineering
Department
Maulana Azad National Institute of
Technology
Bhopal, India

Vipin Singh
Environmental Science and
Engineering Department
Indian Institute of Technology
Bombay
Mumbai, India

Islam I. Teiba
Department of Botany
Faculty of Agriculture
Tanta University
Tanta, Egypt

Vinay Yadav
Indian Institute of Management
Visakhapatnam
Visakhapatnam, India

1

E-Waste Management

Babita Panda, Gyanendra Kumar Panda, and Arjyadhara Pradhan

CONTENTS

DOI: 10.1201/9781003184096-1

1.1 Introduction

At the moment, the industries which make the most profit or are in most demand are electronic industries. In a short span of time, new and advanced technology with up-to-date designs being easier to operate have developed, and consumers have become attracted towards this technology and want to have these advanced operations [1]. In this time of advanced technology, the pace of innovation is quite fast. Advanced and upgraded versions of devices are launched into the market in quick spans of time and offer more, better and advanced functionality than previous versions. Within a few months or a year, masses of people who have the same smart phone model from a brand move to an advanced version of the same model, as shown in Figure 1.1. Due to this rising demand, competition between suppliers for a wider range of models is also increasing [2].

Based on the above trend analysis towards premature obsolescence, it's possible to increased packing density. Other trends which appear in other fields of the electronics sector are transmission speed and improved processors. Product marketing is a well-known example, or we can say factor, that highly motivates consumer intention to upgrade [3–5].

Current fashion is another factor. In the field of smart devices, this is becoming apparent. Nobody anymore wants to be seen carrying a bulky phone with an arial antenna. A few people thought that new models are released too frequently when the old one is not out of date by very much. This results in premature obsolescence and increasing more e-waste. Another factor is

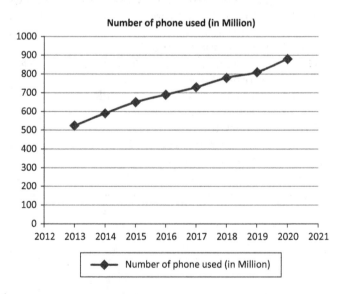

FIGURE 1.1
Uses of different cell-phone models from 2012 to 2022.

the cost of repairing. For instance, it is much easier nowadays for a consumer to buy a new smart phone when the battery life is nearly dead than to invest in a new battery. With so many battery types available in the market for a particular mobile device, consumers do not want to take the risk of buying the wrong type and end up having another problem of completely damaging their currently only troublesome device. Consumer eagerness for new advanced products forces a decrease in the lifespan of existing products, and as a result it increases electrical and electronic waste. An ever-growing portion of the solid waste from cities is due to the increase of e-waste at an accelerated rate [6–8]. Electronic items after use are shipped over oceans, which creates a complicated waste matter consisting of several harmful metals and chemical substances. Consumption of electronic products in India is very high due to the large population.

1.2 What Is E-Waste?

Waste products from electricals and electronics that have been discarded or where the lifespan of the item has ended are known as **electronic waste (e-waste)**. This includes devices that are operated by either electricity or batteries, and includes devices that are thrown away to landfill or given to a reseller organization. Table 1.1 depicts the distribution of e-waste items in different categories [9].

E-waste contains toxins which produce greenhouse gases (GHGs) at the time of their disposal and which impact human health and effect pollution due to low levels of processing. The more the e-waste in landfill, the more these poisonous materials mix into groundwater. A major risk is involved to workers' health and their communities due to the recycling and disposal of e-waste [10–12].

TABLE 1.1

Distribution of E-Waste Items

Home Appliances	Communication and IT Devices	Entertainment Devices	Electronic Items	Office and Medical Equipment
• Microwaves • Home entertainment devices • Electric cookers • Heaters • Fans	• Cell phones • Smart phones • Computer accessories	• Bluetooth Music systems • Televisions • Xbox consoles	• Treadmills • Various electronics equipment	• Copiers/printers • Autoclave • Shredding machines • Fax machines • X-ray equipment • Power supplies • UPS systems • Power distribution systems (PDUs)

1.3 Problems of E-Waste

In a generation of rapid technological advancement, e-waste keeps expanding while smart electronic goods are being invented and introduced to the markets. Electronic devices (smart home devices) can now almost do everything within a short span of time. New models are introduced to the markets even while the current models are working fine [13–15]. The latest versions always provide additional new features that make them superior to the previous ones. New innovators in the field of technology continue to create more smart and advanced electric or electronic devices, designed to make life easier and more convenient in all aspects. Still, consumers seem all too susceptible to quickly upgrading the devices they already have [16–18]. It doesn't matter how satisfied they have been with their devices until now.

1.4 Generation of E-Waste

Mostly the waste is classified in the below categories:

- Household appliances
- IT and communication devices
- Electronic toys
- Electrical utilities
- Health care devices
- Smart devices

Electronics goods which are thrown away by consumers rather than being recycled or reused are known as e-waste [19–20].

1.5 Identification of E-Waste Issues

Several acts are in place for protecting human health and the natural environment from the potential hazards of waste disposal. Laws are there to ensure e-waste management in an environmentally sound manner. An international treaty is in place for reducing hazardous waste movement between nations and to prevent hazardous waste transfer from developed to less-developed countries [21–23]. Through this, the e-waste recycling industry has grown significantly. The recycling industry has been working hard to keep e-waste out of landfill and from being burned in incinerators. The only goal of the recycling industry is to take the reusable parts from unused electronic devices and recycle them for the benefit of other manufacturers. As the mass of electronic waste has increased, so has the recycling industry. Recycling industries today create hundreds of thousands of jobs across the world by recycling electronics products which are no longer in use. Recently, the electronic waste recycling system has been implemented [24–26].

1.6 Environmental Impact of E-Waste

It was estimated that the global production of e-waste in 2020–2021 was more than 50 million tons. In India, like in other developing countries, e-waste management is dominated by the informal sectors which are not monitored by any form of government, according to a survey which estimates that more than 90% of e-waste is being processed in this sector [27–30]. India is one of the top five e-waste producing countries in the world, with an estimated annual production of two million tons. There is a risk involved to health from direct contact with toxic materials that come from e-waste, according to the World Health Organization (WHO).

Risks can be in the form of:

- Intake of the toxic fumes
- Gathering of chemicals in soil, water, and food

E-waste risks are not only bound to people but land and sea animals as well. The risks are exceptionally high for developing countries because some developed countries send their e-waste to those developing countries [31–33]. Both the people that work with e-waste and the people that live around it can suffer from the same detrimental effects. The atmosphere, hydrosphere, lithosphere, and biosphere are adversely impacted because of e-waste. For the manufacturing of electronics, mines for extracting metals are used, some of which are now totally depleted. Because of this, there is now a shortage of metals. Entire ecosystems are now in danger because of mining operations that explode mines in areas such as forests, thereby destroying the habitat of plants and animals living there [34].

One of the major impacts to the environment from e-waste is global warming. A study indicated that about 2% of CO_2 produced and discharged in the atmosphere comes from IT industries. CO_2 is generated from IT devices at the time of their mining, manufacture, usage, and their disposal as e-waste that is dumped into landfill [35]. The production and discharge of CO_2 from IT industries is called cyber warming. Old landfill sites with these uncontrolled dumps may turn into a toxic bomb due to the production and discharge of

TABLE 1.2

Toxic Materials Present in E-Waste and Their Effects

Material	Occurrence	Effect
Lead	Batteries, crystal, glass	The toxic material present in lead affects kidney, the reproductive system, as well as mental growth of children
Chromium	Data tapes	Damage of liver, kidney; can cause bronchitis
Mercury	Fluorescent lamps and batteries	Affects the central nervous system, kidney and digestive systems
Cadmium	Batteries, computer accessories, and cathode ray tubes (CRTs)	Cadmium causes itai-itai diseases
Lithium	Batteries	May become explosive when wet
Nickel	Batteries and cathode ray tubes (CRTs)	May cause allergic reactions
Barium	Semiconductors, diodes, PV cells	May explode when wet
Rare earth elements	CRT screen in fluorescent lamps	Irritates skin and eyes
Arsenic	Semiconductors, diodes, LEDs, solar cells	Causes long-term problems to health

excess CO_2. Another gas which produces and discharges from these landfill sites and affects global warming is methane gas. Due to global warming, ice melts into water and increases the water level in oceans, thus causing the ocean levels to rise, leading to flooding [36–38]. A study says that a reduction in Arctic ice causes the removal of the permanently frozen layer on or under the earth's surface, meaning the ocean's absorption of CO_2 is further reduced, which leads to further warming. Due to the disturbance in ocean ice layers, entire life cycles will change and impact the environment. If we do not take action regarding this, then all ecosystems will be destroyed. Table 1.2 categorizes the different toxic materials present in different forms of e-waste as well as their effects.

We need to put proper recycling processes for this e-waste in place to protect ourselves and future generations. In considering these calamities, deregulation acts have been introduced by several progressive countries over the past few decades.

1.7 Proper Disposal of E-Waste

Because consumers will not stop buying new devices, it is important to strengthen the message to recycle old electronic models and not to simply throw them out. There are major risks involved to the environment if electronics devices are thrown into landfill. E-waste recycling provides many benefits to our environment [39]. A good solution would be to hand over those unused electronic devices to legitimate firms that have experience of performing environmentally friendly recycling of electronic products. By reusing, refurbishing, and recycling e-waste, we can save our environment from the damage caused by it. Green technologies are in trend now, and companies are being created in this sector to keep our environment clean and pollution free. Lastly, whether products are made using green technology or not, there is a need to opt for proper scientific recycling methods because even these green products can cause environmental problems [40].

1.8 E-Waste Management

In order to manage e-waste, many developed and developing countries have created some regulations and taken some initiatives. This will be discussed in this section and given in Figure 1.2.

- Extended producer responsibility (EPR) and e-waste
- Public schemes for e-waste management
- Problems and threats for policy execution
- Designing of a powerful e-waste management system

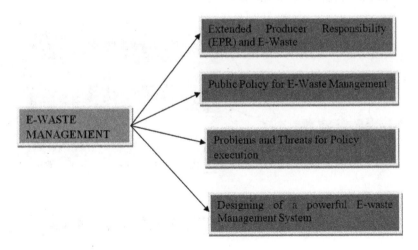

FIGURE 1.2
The different strategies of e-waste management.

1.8.1 Extended Producer Responsibility (EPR) and E-Waste

For regulating e-waste everywhere, the EPR concept is popularly used and points out producers' accountability for the end-of-life execution of stocks [41–43]. According to the EPR concept, the producer has to bear the costs related to the end-of-life recycling of their brands. The Organization for Economic Co-operation and Development (OECD) focuses on two extensive intentions of the EPR scheme. The primary intention is to transfer the responsibility of administration of e-waste from local governments to manufacturers. The other intention is to force manufacturers to consider the environmental impact at the time of disposal of their product. The EPR does this by providing incentives to producers if they consider the environmental issues while designing their product [44–46]. In this way, producers are motivated to use materials in their products that are more recyclable or less toxic so that they get monetary incentives for designing their products under this EPR concept.

Under the EPR scheme, manufacturers can be made accountable in four specific ways as shown in Figure 1.3.

- Economic responsibility: for proper disposal of their product, the manufacturer is compensated in terms of tax against the costs of e-waste processing [47]
- Physical responsibility: collecting the products back from customers at the end of the product's life. The stock take-back fulfillment may impose the collection of rate targets
- Information responsibility: this includes making data associated with the brands available, along with such essential information as manufacturer specifications

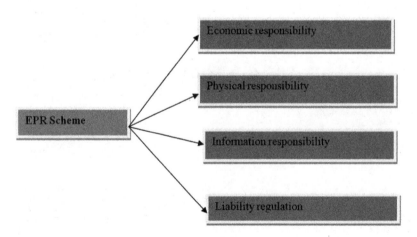

FIGURE 1.3
Four specific ways of implementation of the EPR scheme.

- Liability regulation: this indicates the financial accountability for environmental pollution and its recovery

1.8.2 Public Schemes for E-Waste Management

In 2011, India's first e-waste policies come into reality, known as e-waste rules. These rules were set up in 2011 and used the EPR approach for the management and treatment of e-waste. With this approach, producers are required to set up collection centers, coming under physical responsibility and instructions to the consumers about product use, and depositing the product at collection centers, coming comes under information responsibility [48–50]. However, these rules created a need for formal dismantling and recycling centers, which did not effectively improve the situation. Those regulations were not properly framed as there were no take back targets for the manufactures. That led to a reformation of these rules in 2016 and then again in 2018, which enforces producers to take back the certain percentage of their products taken in the last year.

1.8.2.1 Main Highlights of the E-Waste Rules

- As per the new rules reformed in 2018, the assembly, repository, shipping, recovery, and disposition of e-waste shall be done according to the guidelines. According to the board of the Environment Safety Act, the manufacturer, producer, importer, transporter, and recycler shall be liable to pay compensation if they fail to follow the guidelines [51–53].
- As per the current modifications of the rules, the Central Pollution Control Board (CPCB) may organize unplanned inspections of materials located in the retail unit to audit and confirm the decrease of poisonous materials. The government should bear all costs for sampling and testing, something which was not there in the earlier rule: in earlier cases, the costs were carried by the producers [54].

1.8.2.2 Duties of Different Shareholders

- **Manufacturer responsibility**
  - During the manufacturing of any product, there is the possibility of generation of e-waste. It is the duty of the manufacturers to assemble this e-waste and process it for disposal
  - The storing and shipping of e-waste should not cause any damage to the environment: this is also the responsibility of the manufacturers
  - The duty of State Pollution Control Board is to audit and check all the records which should be maintained by the manufactures about the generation and recycling of e-waste [55]

- **Producer responsibility**
  - For enforcing extended producer responsibility, the producer of electrical and electronic equipment shall be responsible for the given structure, namely, assemblage and processing of e-waste produced from the expired products as well as old waste obtained from the date on which the rules came into force. The system used for processing of e-waste from expired products consists of collecting them from their collection centers and sending them to certified disposal units as per the rule.
  - For certain materials like fluorescent lamps or other lamps which contain mercury, simple recycling is not possible. In those cases, a pre-treatment is required to exhaust the mercury and decrease the quantity of waste to be recycled. Proper channelization is required from service center to the treatment, Storage and Disposal Facility centre.

This brings about the installation of many advanced recycling and dismantling sections, properly designated with certified jurisdiction. These reformations of the rules have initiated take-back targets for producers, whereby manufacturers need to take back a specific share of their items depleted in the past year. The take-back shares will increase from 10% in the year 2017–2018 to 70% from 2023.

Because of the reformation of the rules, in past years there have been several changes in the Indian e-waste sector, for example:

- The producers taking a part of the burden
- Development of the certified waste treatment sector
- Formation of producer responsibility organizations (PROs)
- Building environmentally friendly product
- Retrieving multiple kinds of e-waste

Despite all these developments, management of e-waste is a still great challenge because most e-waste is processed informally in India. The role of the informal sector is predominant not only in the collection of e-waste but also in recovery and recycling. The large population of India, mainly marginalized parts of society, depends on this sector. Unknowingly, they are using an unreliable method that is harmful to their health, and in turn it is a threat to the environment and health of society at large.

1.8.3 Problems and Threats for Policy Execution

1.8.3.1 Less Knowledge on E-Waste Formation Rates

The e-waste regulations policy made in 2012 gave all accountability to the state pollution control boards (SPCBs) for installation of new recycling and dismantling units. After many years, the SPCBs have not issued inventories as yet. There have been no proper statistics developed until now for the estimation of quantities of e-waste generation. The sales information about their brands, which is a critical calculation in the assessment of the amount of e-waste, is usually accessible at an aggregation of the national level, making it difficult to know the supply at the state levels. Along with that, there is an illegal supply of e-waste from other countries too. It is difficult for the SPCBs to get all this information, and the implementation of proper, transparent system is needed. There is much less information available about the quantities of e-waste imported into the country. Modeling systems for precise assembly, transportation, and processing needs the correct information regarding waste production, combination, and outflows.

1.8.3.2 Informal Organizations' E-Waste Management Practices

There is no facility offered by the formal sector for taking back and processing of e-waste from consumers. Therefore, the amount of waste handled in the formal sector remains very low. As the formal sector does not generate very much waste, most facilities operate below their rated capacities. On the other hand, the informal sector has better strategies than the formal sector to encourage customers to bring back their waste. A lack of awareness regarding the threat of the material which e-waste contains and the waste management practices and stance of informal organizations are a major threat to their employees and others as well. It is high time this issue is resolved and a robust system for e-waste management is developed [56].

1.8.3.3 Poor Regulation and Enforcement

In the 2012 regulations, there were many shortfalls. There was no mandatory take-back system for producers, which was then addressed in the 2016 amendments. In the 2016 amendments, it was clearly mentioned that without having collection targets, no incentives will be provided to producers. In addition to this, the reformation provided more positive regulatory measures by determining continuous and more rigid collection targets. As well as this, this regulatory endeavor gave a specific task to the previously unsuitable regulatory agencies. The role of the regulatory agencies is to audit all e-waste management processes followed by the producers and then grant authorization. But because of poor enforcement, a lack of clarity, and reluctance to openly share the data on actions, natural policy execution in India has been affected, which is a threat to the future of the management of e-waste.

1.8.4 Designing of a Powerful E-Waste Management System

The regulations can be made effective by constant evaluation and monitoring as well as the regulator having to bring in necessary changes to the regulations. In addition, the government may have to show relevant performance in relation to the different shareholders in the structure. Some important points are discussed next.

1.8.4.1 Informal Sector

For the designing of a robust system, the first step would be to specifically identify the informal sector as a relevant shareholder in any forthcoming e-waste system. But the management methods of this sector pose severe environmental and health threats to workers. First of all, the problems and challenges of the informal sector need to be addressed. A crucial step would be to involve their employees so as to understand their right to a livelihood, and to develop understanding and realization of their problems along with recognizing potential solutions. The government should constitute a forum consisting of various stakeholders like the informal sector, NGOs engaged with them, PROs, and certified disposers and manufacturers under the MoEFCC at the central level as well as at the state level [57]. The objective of such forums should be to define roles for each stakeholder and work towards achieving them.

1.8.4.2 Policy Reformation under EPR

Under the EPR scheme, the ministry should re-evaluate the policy instruments. The assembling of e-waste from service centers with the help of the informal sector and a compulsory take-back with collection targets may not be the optimal instrument. In addition, there are several responsibilities of the producer. For a certain percentage of quantity purchased, an advanced disposal fee (ADF) could be collected, and the producers would be relieved from the physical responsibility of the collection of e-waste from service centers. In this way, revenues produced may be used in different ways.
Some possibilities include:

(a) Incentives offered to customers to submit their e-waste at service centers
(b) Funds to be allocated to certified disposers
(c) Provision of skill development training to assist informal sector workers

This selection can be made within the institute forum suggested by the informal sector. Determination of the right fee is the main issue with economic instruments. According to the principles of economics, the right fee would be equal to the minimal peripheral cost of the end-of-life product. In practice, the estimation of peripheral cost is difficult to calculate, so the price should be sufficient to fund robust, environmentally safe e-waste processing and disposal. This would also include the subsidies for designing the changes in the product to make it more environmentally safe, which has been one of the main aims of the EPR approach universally. The policy structure needs to target the improvement of advanced systems and/or system exchange to boost an all-over spread of the use of environmentally safe e-waste disposal technologies.

1.8.4.3 Execution of Regulations

By adopting economic instruments such as an advanced disposal fee (ADF), producers would be relieved from the regulatory burden. The funds generated from ADFs would be spent on technologies to make the tax collection process easy. The SPCBs and the CPCB will still be needed to direct and execute the process, with rules made for collection centers, dismantlers, recyclers, and PROs. The policies made by MoEFCC must ensure that the e-waste management system is clear. All steps taken, as well as the related documents, must be publicly available on the websites of the SPCBs. This data needs to be updated regularly and publicly available; this is required in the execution of policies.

1.8.4.4 E-Waste Imports

In the 2016 amendments, it is clearly mentioned that the import of e-waste is allowed for reuse and recycling but not for final disposal. In a country like India, there is no proper infrastructure for recycling of e-waste, so all kinds of imports should be banned.

1.8.4.5 Public Awareness

Public awareness is the most important criterion in every aspect. The recent e-waste regulations need the producers to maintain a proper database on their websites, like information on the effects of e-waste, proper disposal methods, and other challenges. Some awareness programs can also be useful to the public if run at regular intervals. In India, the conducting of such awareness programs among the public is very low, because of which overall awareness levels remain low. So the rigid instruction to producers to conduct these programs at regular intervals may improve the situation. Simultaneously, producers should be compelled to organize the programs through the organizations engaged in different areas. There may be collaborations with other programs, like municipal solid waste programs, initiated by the government. These awareness campaigns should have the objective of achieving the secure use of e-waste but also decreasing their use. The evaluation could be done by the government, depending on the quality and quantity of the awareness campaigns held by the producer and the outcome achieved.

1.9 Treatment of E-Waste

E-waste is considered the most complex among a wide range of different kinds of waste. There are multiple forms of e-waste, so various technologies are required to cover the wide range of products. Consequently, the employment of various technologies varies for different regions and countries. A robust management system is required for the assembling and proper treatment of e-waste, which consists of different operators as well as significant logistical and technical assets. To save our environment from damage caused by this e-waste, many researchers have considered factors like the attitude of customers toward recycling, the local government context, and the position of the producer toward this aspect. Reusing e-waste is a preferable solution from the aspect of environmental concerns. E-waste reuse is the most preferable solution, as the disassembly and recycling as well as disposal of e-waste may be harmful and a threat to the environment and health. The recovery and recycling of these products is a very important action towards the contribution to sustainable development. Therefore, the precise legislating, design, and execution of this waste is environmentally important and needs to be solved at all levels. The poisonous material from the erosion and disintegration of e-waste may discharge an amount of harmful substances in the air, in the form of solids and liquids, which affects the health of humans, animals, and plants.

1.10 Conclusion

In 2007, according to the UN Environmental Programme (UNEP), 439,000 units of e-waste were generated, among which the computer e-waste alone was 56,300. In 2010 in India, units of computers installed were about 47,000,000. As time progresses, e-waste generation also increases. These computers generate a vast amount of CO_2. To save the environment from CO_2 emissions, immediate action should be taken. A computer produces about 675 kg of CO_2 when used for 24 hours. Improper e-waste disposal causes damage to the earth. The solution to reducing CO_2 emissions is by farming trees. Tree planting can support environmental repair from these damages. In a year, one tree consumes about 1.3–6.8 kg of CO_2. So the generation of CO_2 from one computer means almost 100–500 trees would be needed to recover from that damage.

Especially in developing countries, the increase in the use of electronic products over recent years and the corresponding increase in e-waste cause specific environmental issues for authorities. From the aspect of e-waste management, India has faced many challenges because of the limited impact of the old policies. This article focuses on the challenges that India faces like informal organizations e-waste management methods: poor policy design and execution, and lack of public awareness. Various shareholders must be engaged meaningfully to build a powerful e-waste management system of the future.

In a country like India, the informal sector has great importance in managing e-waste. They have the strength to collect waste from all areas of the country. They have the ability to collect and dismantle waste from the urban poor. The issue remains to find the right connection between the law and the informal sector. If the government acknowledges the importance and contribution of this sector, only then is this possible. The MoEFCC plays an important role here in defining the place of informal organizations in e-waste management. They need to review the policies more precisely to identify the locations of informal organizations. In creating public awareness, the government has a great role to play, which could be an important step in altering the behavior of consumers. Awareness campaigns could be held along with other programs which would increase knowledge about the management of e-waste.

References

1. *A New Circular Vision for Electronics Time for a Global Reboot, A New Circular Vision for Electronics: Time for a Global Reboot*, World Economic Forum, 2019.
2. S. Sthiannopkao, M.H. Wong, Handling e-waste in developed and developing countries: initiatives, practices, and consequences, *Science of the Total Environment*, 463–464 (2013), 1147–1153.
3. D. Yang, Y. Chu, J. Wang, M. Chen, J. Shu, F. Xiu, Z. Xu, S. Sun, S. Chen, Completely separating metals and nonmetals from waste printed circuit boards by slurry electrolysis, *Separation and Purification Technology*, 205 (2018), 302–307.
4. L.H. Yamane, V.T.D. Moraes, D.C.R. Espinosa, J.A.S. Tenório, Recycling of WEEE: characterization of spent printed circuit boards from mobile phones and computers, *Waste Management*, 31 (2011), 2553–2558.
5. X. Wang, Y. Guo, J. Liu, Q. Qiao, J. Liang, PVC-based composite material containing recycled non-metallic printed circuit board (PCB) powders, *Journal of Environmental Management*, 91 (2010), 2505–2510.
6. D. N. Perkins, M. Drisse, T. Nxele, P. D. Sly, E-Waste: a global hazard, 2014 Icahn School of Medicine at Mount Sinai, *Annals of Global Health*, 80 (2014), 286–295.
7. K. Grant, F. Goldizen, P.D. Sly, M. Brune, M. Neira, M. Berg, R. Norman, Health consequences of exposure to e-waste: a systematic review, *The Lancet Global Health*, 1 (December 2013), 350–361.
8. J. Willner, Extraction of metals from electric waste by bacterial leaching, *Environment Protect Ion Engineering Journal*, 39 (2013), 197–207.
9. A. Khaliq, Metal extraction process for electronic waste and existing industrial routes: a review and Australian perspective, *Resources* 3 (2014), 152–179.

10. E. Ma, *Recovery of Waste Printed Circuit Boards Through Pyrometallurgy, Electronic Waste Management and Treatment Technology*, pp. 247–267, 2019.
11. R. Sinha, G. Chauhan, A novel eco-friendly hybrid approach for recovery and reuse of copper from electronic waste, *Journal of Environmental Chemical Engineering*, 6 (2018), 1053–1061.
12. Y. Chu, M. Mengjun Chen, S. Chen, B. Wang, K. Fu, H. Haiyan Chen, Micro-copper powders recovered from waste printed circuit boards by electrolysis, *Hydrometallurgy*, 156 (2015), 52–157.
13. L. Dascalescu, T. Zeghloul, A. Iuga, Electrostatic separation of metals and plastics from waste electrical and electronic equipment, *WEEE Recycling*, 1 (2016), 75–106.
14. R. Jujuna, Q. Yiminga, X. Zhenming, Environment-friendly technology for recovering nonferrous metals from e-waste: Eddy current separation, *Resources, Conservation and Recycling*, 87 (2014), 109–116.
15. M. Kaya, Recovery of metals and nonmetals from electronic waste by physical and chemical recycling processes, *Waste Management*, 57 (2016), 64–90.
16. X. Yi, Y. Qi, F. Li, J. Shu, Z. Sun, S. Sun, M. Chen, S. Pu, Effect of electrolyte reuse on metal recovery from waste CPU slots by slurry electrolysis, *Waste Management*, 95 (2019), 370–376.
17. A.V. Kavitha, Extraction of precious metals from E-waste, *Journal of Chemical and Pharmaceutical Sciences*, Special Issue 3 (October 2014), 147–149.
18. T. P. Bezerraa, A.V. Grillob, B.F. Santos, Optimization of the copper recovery from printed circuit boards using artificial intelligence as strategy, *Chemical Engineering Transactions*, 74 (2019), 1231–1236.
19. V. Flores, B. Keith, C. Leiva, Using artificial intelligence techniques to improve the prediction of copper recovery by leaching, *Journal of Sensors*, (2020), 2454875.
20. R. Widmer, H. Oswald-Krapf, D. Sinha-Khetriwal, M. Schnellmann, H. Böni, Global perspectives on e-waste, *Environmental Impact Assessment Review*, 25 (2005), 436–458.
21. P. Huang, X. Zhang, X. Deng, Survey and analysis of public environmental awareness and performance in Ningbo, China: a case study on household electrical and electronic equipment, *Journal of Cleaner Production*, 14 (2006), 1635–1643.
22. I. C. Nnorom, O. Osibanjo, Overview of electronic waste (e-waste) management practices and legislations, and their poor applications in the developing countries, *Resources, Conservation and Recycling*, 52 (2008), 843–858.
23. Z. Wang, B. Zhang, J. Yin, X. Zhang, Willingness and behavior towards e-waste recycling for residents in Beijing city, China, *Journal of Cleaner Production*, 19 (2011), 977–984.
24. G. Davis, S. Heart, Electronic waste: the local government perspective in Queensland, Austria, *Resources, Conservation and Recycling*, 52 (2008), 1031–1039.
25. J.M. Yoo, J. Jeong, K. Yoo, J.C. Lee, W. Kim, Enrichment of the metallic components from waste printed circuit boards by a mechanical separation process using a stamp mill, *Waste Management*, 29 (2009), 1132–1137.
26. I.C. Nnorom, O. Osibanjo, Electronic waste (e-waste): material flows and management practices in Nigeria, *Waste Management*, 28 (2008), 1472–1479.
27. I.C. Nnorom, O. Osibanjo, Toxicity characterization of waste mobile phone plastics, *Journal of Hazardous Materials*, 161 (2009), 183–188.
28. C.F. Shen, S.B. Huang, Z.J. Wang, M. Qiao, X.J. Tang, C.N. Yu, D.Z.Shi, Y.F. Zhu, J.Y. Shi, X.C. Chen, K. Setty, Y.X. Chen, Identification of Ah receptor agonists

in soil of E-waste recycling sites from Taizhou area in China, *Environmental Science & Technology*, 42 (2008), 49–55.

29. X.B. Liu, M. Tanaka, Y. Matsui, Economic evaluation of optional recycling processes for waste electronic home appliances, *Journal of Cleaner Production, N*, 17 (2009), 53–60.

30. M.H. Wong, S.C. Wu, W.J. Deng, X.Z. Yu, Q. Luo, A. Leung, C. Wong, W.J. Luksemburg, A.S. Wong, Export of toxic chemicals: a review of the case of uncontrolled electronic-waste recycling, *Environmental Pollution*, 149 (2007), 131–140.

31. J.R. Cui, E. Forssberg, Mechanical recycling of waste electric and electronic equipment: a review, *Journal of Hazardous Materials*, 99 (2003), 243–263.

32. B.Y. Wu, Y.C. Chan, A. Middendorf, X. Gu, H.W. Zhong, Assessment of toxicity potential of metallic elements in discarded electronics: a case study of mobile phones in China, *Journal of Environmental Sciences-China*, 20 (2008), 1403–1408.

33. S. Herat, Recycling of cathode ray tubes (CRTs) in electronic waste, *Clean-soil Air Water*, 36 (2008), 19–24.

34. S. Herat, Sustainable management of electronic waste (e-waste), *Clean soil Air Water*, 35 (2007), 305–310.

35. J. Williams, L.H. Shu, Analysis of remanufacturer waste streams across product sectors, *Cirp Annals-Manufacturing Technology*, 50 (2001), 101–104.

36. J.R. Cui, E. Forssberg, Mechanical recycling of waste electric and electronic equipment: a review, *Journal of Hazardous Materials*, 99 (2003), 243–263.

37. J.R. Cui, L.F. Zhang, Metallurgical recovery of metals from electronic waste: a review, *Journal of Hazardous Materials*, 158 (2008), 228–256.

38. W. Kiatkittipong, P. Wongsuchoto, K. Meevasana, P. Pavasant, When to buy new electrical/electronic products?, *Journal of Cleaner Production*, 6 (2008), 1339–1345.

39. P. Matkowski, K. Friedel, J. Felba, Treatment of waste electrical and electronic equipment. Technologies of disassembly, recycling stages, *Progress in Eco-Electronics*, Tele Radio Research Institute, pp. 143–147, 2008.

40. P. Matkowski, K. Friedel, J. Felba, System of waste electrical and electronic equipment management – potential directions of development in a few-year perspective, *Progress in Eco-Electronics*, Tele Radio Research Institute, pp. 178–184, 2008.

41. R. Gadh, H. Srinivasan, S. Nuggehalli and R. Figueroa, Virtual disassembly-a software tool for developing product dismantling and maintenance systems, Annual Reliability and Maintainability Symposium. 1998 Proceedings. International Symposium on Product Quality and Integrity, 1998, pp. 120–125, Anaheim, CA, USA.

42. E. Zussman, M. Zhou, R. Caudill, Disassembly PetriNet approach to modeling and planning disassembly processes of electronic products, Proceedings of the IEEE International Symposium for Electronics and the Environment, Oak Brook, IL, pp. 331–338, 1998.

43. K.E. Moore, A. Gungor, S.M. Gupta, Disassembly process planning using petri nets. Proceedings of IEEE International Symposium on Electronics and the Environment, Oak Brook, IL, pp. 88–93, 1998.

44. K.E. Moore, A. Gungor, S. M. Gupta, Disassembly petri net generation in the presence of XOR precedence relationships, Proceedings of the IEEE International Conference on Systems, Man, and Cybernetics (SMC'98), 11–14 October 1998, San Diego, USA, pp. 13–18, October 1998.

45. C. T. Kuo, H. C. Zhang, S. H. Huang. Disassembly analysis for electromechanical products: a graph-based heuristic approach. *International Journal of Production Research*, 38(5) (2000), 993–1007.

46. S.M. Gupta, P. Veerakamolmal, A case-based reasoning approach for the optimal planning of disassembly process, *Proceedings of the International Seminar on Reuse*, 141 (1999), 150.

47. K.-K. Seo, J.-H. Park, D.-S. Jang, Optimal disassembly sequence using genetic algorithms considering economic and environmental aspects, *International Journal of Advanced Manufacturing Technology*, 18(371) (2001), 380.

48. C. Xueyuan, H. Yujin, H. Junjun, L. Chenggang, Decision on the optimum route of disassembly in the process of recycling product, *Journal of Huazhong University Of Science & Technology*, 28(3) (2000), 27–29.

49. R.W. Chen et al., *Product Design for Recycle Ability: A Cost Benefit Analysis Model and its Application*, IEEE 8 (1993), 178–183.

50. Z.F. Liu, X.P. Liu, S.W. Wang, G.F. Liu, Recycling strategy and a recyclability assessment model based on an artificial neural network, *Journal of Materials Processing Technology*, 129 (2002), 500–506.

51. L. Zhifeng, L. Guangfu, L. Jiuguang, Z. Huabin, J. Jibing, J. Zhongwei, Study on decision-making for recycling process of used products, *Computer Integrated Manufacturing Systems*, 11(8) (2002), 876–880.

52. M.S. Sodhi, B. Reimer, Models for recycling electronics end-of-life products, *OR Spektrum*, 23 (2001), 97–115.

53. W.S. Wang, L. Zhifeng, L. Guangfu, P. Xiao Yong, Material recycling process planning based on agent, *Computer Integrated Manufacturing Systems (CIMS)*, 9(11) (2003), 1001–1005.

54. Z. Jie-han, X. Guang-leng, Z. He-ming, L. Hu., Y. Shu-Zp, Product life cycle engineering for environment, *Manufacture Automation*, 24 (4) (2002), 30–33.

55. Y. Y. Cao, G. F. Liu, Z. F. Liu, C.X. Pan, Analysis of life circle of green products based on petri web, *Industrial Engineering Journal*, 5(4) (2002), 31–35.

56. C. Xueyuan, H. Yujin, H.J. Chenggang, Decision on the optimum route of disassembly in the process of recycling product, *Journal of Huazhong University of Science & Technology*, 28(3) (2000), 27–29.

57. Z.F. Liu, X.P. Liu, S.W. Wang, G.F. Liu, Recycling strategy and a recycle ability assessment model based on an artificial neural network, *Journal of Materials Processing Technology*, 129 (2002), 500–506.

2

Waste Prevention: Its Impact and Analysis

**Arjyadhara Pradhan, Sarita Samal, Babita Panda,
and Biswaranjan Acharya**

CONTENTS

2.1 Introduction

One of the major environmental issues today is waste. While much work is put into gathering and recovering waste materials. From an environmental standpoint, waste management is preferable to other forms of waste treatment (landfill, energy recovery, and recycling), because the manufacture and disposal of waste materials and substances is to be avoided. Waste prevention (Directive 2008/88/EC) is characterized as actions taken before a substance, material, or product becomes waste. It entails strict waste avoidance, i.e., waste generation reduction. The term also refers to extending the life of goods and reusing them, as well as the qualitative element of minimizing waste's hazardousness. Waste prevention is something that can happen in any situation. As a result, the regulatory structure regulating waste prevention encompasses a diverse set of directives and regulations, such as REACH (Regulation (EC) No. 1907/2006) and the Eco-design Directive (Directive 2009/125/EC), both of which seek to avoid creating waste during the product design process. This makes sense because waste can't be avoided once it's been generated. So far, waste management has been regarded as a policy area distinct from waste prevention [1]. End-of-life options, such as collection, landfill, incineration, and recycling, are essential parts of waste treatment [2]. As a result, avoidance of creating the waste in the first place played no part in local waste management so far. One of the amendments made to the European Waste Framework Directive in 2008 was a greater emphasis on waste reduction (EU, 2008). Despite the fact that waste reduction should be prioritized, it is seldom incorporated into local waste management [3].

The modification of the Directive resulted in member states being required to establish national waste management strategies. Since waste prevention can occur at any point in the value chain, it is advantageous to establish comprehensive national strategies. However, until now, national prevention policies have seldom assigned specific tasks to specific actors [4], and local waste managers (LWMs) are rarely mentioned as responsible actors. As a result, LWMs are rarely required to include waste reduction in their waste management plans. In addition to their national waste management strategies, member states must recycle at least half of the waste produced by their households (EU, 2008). This necessitates improvements to the waste management system's socio-technical system, such as the implementation of collection systems, logistics, treatment plants, and marketing and sales of the secondary raw materials. Since these processes fall under the conventional domain of local waste management, LWMs play a key role in these infrastructural innovations. Nonetheless, these modifications are last-ditch efforts. As with other environmental issues, it's better to go after the source of the problem rather than addressing the symptoms once they've appeared. As a result, the primary goal should be to reduce overall waste quantities. Otherwise, society risks investing in waste facilities that would not have been needed if preventive measures had been implemented earlier [5]. As a result, before making long-term investments in treatment facilities, LWMs should consider what can be done to avoid waste in their environment. To put it another way, more integrated resource management is needed. The fact that waste reduction and reuse are outside the trajectories of local waste management is an impediment to integrating waste prevention into local

waste management. As a result, local waste actors lack experience, awareness, and skills in waste prevention [6].

Waste prevention has received more attention in the scientific community in recent years, and this awareness could help LWMs get started with
local waste prevention. As a result, we present a summary of the scientific
evidence on waste reduction in this chapter, which has been chosen for its
importance to LWMs. The ability of LWMs to influence waste generation
through the supply chain is restricted to the final stages of a product's life
cycle. Local waste management programs would therefore primarily target homes, with some retail and industry thrown in for good measure. The
essential data whereupon city waste can be calculated is based on the sum
of family waste produced [7]. The fundamental data required is first to survey the potential and importance of waste decrease, and second to focus on
center regions, for example, waste streams, to help prediction of waste. In the
UK, family waste can comprehensively be diminished by 0.5–1 kg of waste
per family per week, for example, through appropriate campaigns [8, 9]. In
addition, there is the belief that the most probable areas with a potential to
reduce waste are firstly food waste and secondly paper waste. The principal
reason is that people in families can more easily affect the levels of household waste generated, like avoiding using plastic packaging for example. The
genuine potential relies upon the amount of waste and synthesis locally, and
will subsequently contrast between nations. Reference [10] proposes a 20%
decrease of both food (3.3 kg–1 kg every year) and paper waste (3.7 kg–1 kg
every year) as possible in Austria; while Reference [11] recommends that 34%
of food waste in Swedish families is avoidable. This is equivalent to 58 kg
per family each year. Gentil et al. [12] evaluate the natural effect potential
of 20%, and presume that forestalling food and paper squander has a high
natural effect inferable from the upstream impacts of kept away from creation. The reason is that anticipation and avoidance of food waste has a
higher ecological effect than any treatment choice, for example, incineration
and anaerobic assimilation, would have. Subsequently, waste anticipation of
these two waste streams could play a critical part in asset protection and
environmental change (contingent upon the public waste framework as well
as financial considerations). As indicated by the definition, reuse is thought
of as the avoidance of waste generation. In view of contextual investigations,
Reference [13] gauges that reuse of cumbersome waste, like furnishings,
could be expanded altogether in the UK from 2–3% to 40%. References [14–
16] likewise consider reuse as an approach to forestall the creation of waste;
however, the authors don't predict the potential quantitatively. A quantitative appraisal would rely intensely upon the qualities of the waste.

As a rule, reuse could allow ecological benefits if reused items were used
instead of obtaining new items; however, no relevant peer-explored considerations have been found that look at reusing items as a substitute to obtaining new items [17]. Besides this, reuse may really fuel utilization, as the cash
saved by purchasing recycled products could be utilized for obtaining new

items, as proposed by [15, 18], where evaluate the asset protection capability of the reuse of electrical machines to be doing 33%, yet to be immaterial, in any case, contrasted and productive reusing, which offers more to asset protection.

Thus, they suggest that different avenues other than asset preservation should be considered to add to the choices available. The ecological effects of reuse are consequently generally unseen, and future endeavors to evaluate them require a sensible interpretation of framework limits. There may likewise be an impressive social potential in including waste avoidance and reuse in nearby waste administration centers. Social points of view, as stressed by [14], incorporate the way that reuse and second-hand merchandise give individuals from low-income families the capacity to maintain expectations for everyday comforts at moderate costs. Besides this, this reuse should be relied upon to supplant utilization of new things, subsequently having natural advantages. Local work creation and consequently the re-employment of unemployed individuals are additionally underlined by [19, 20], who additionally recommend that there is a financial potential in reuse as well. Reuse may accordingly be able to add to each of the three measurements of manageability.

2.2 Waste: A Brief Idea

Waste is one of the major environmental issues that have to be seen as a top priority. Many things which are discarded and rejected are those items which we consider not useful. There is a difference between waste and by-products. By-products are something that can be processed and integrated into some useful product, whereas waste is something that has been totally discarded. The importance lies in realizing the levels of actual waste components in a product which we consider as waste. There are basically two types of waste, i.e., controlled waste and non-controlled waste. The waste generated from domestic areas like household and municipal solid waste comes under the category of controlled waste. Even waste from construction sites, industries, and other commercial premises falls under this category. Waste generated from agriculture, mining activities, quarries, and other such operations is known as non-controlled waste.

Municipal solid waste (MSW) includes paper, cardboard, glass, plastics, textiles, metals, and wood waste. Agricultural waste includes manure from farms, slurry, silage effluent, cereal and crop residues, and rotten output. Among these different types of waste, some are hazardous, as they contain toxic and explosives. Oily waste and waste from chemical plants are mostly dangerous, as they create environmental pollution. Many organic processes give out strong chemicals as output, which cause air and water pollution.

2.2.1 Impacts of Waste

Waste is harmful if it exceeds a certain limit and greatly impacts society. The various impacts of waste are shown in Figure 2.1. The following impacts will now be discussed:

1. Impact on environment
2. Impact of health
3. Impact on society

2.2.1.1 *Environmental Impact*

Waste disposal causes various environmental problems. Different types of waste are released into the environment in various ways; some are deposited in landfills, some are dumped, some are put in quarries, and some are buried inside holes in the ground. But very often in India we find waste stacked and gathered by the side of roadways without proper disposal. This waste eventually rots and produces a very bad smell; sometimes it also generates methane gas which is explosive in nature and affects the ozone layer. One type of pollution caused by these waste disposals is leachate. This is produced during waste decomposition processes. Due to the foul smell of waste and dumps, vermin and even different kinds of infectious insects become attracted to these sites. Leachate is also easily spread in the atmosphere.

Studies have shown that waste incineration processes produce various toxic substances, for example dioxins are produced when plastic is burnt.

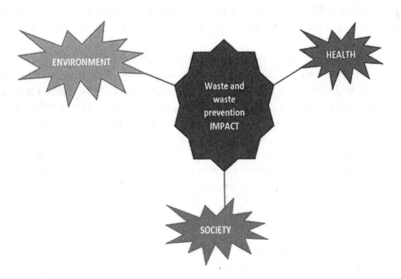

FIGURE 2.1
The impact of waste and waste prevention on various fields.

Chemical incineration processes release gases which cause air pollution. Eventually, air pollution leads to acid rain, and decay and damage to buildings. Some other environmental effects are the corrosion of metal structures, buildings, and pipes.

2.2.1.2 Health Impact

Pollution and waste are always a source of health problems. Different kinds of health hazards can be observed when environmental pollution increases. It is mostly workers involved in waste management who are highly prone to the effects of different harmful and toxic substances, thereby creating difficulties in leading a healthy life. Air pollution caused due to the burning of waste in dumps causes asthma and lung problems. Sometimes waterlogging and oil spills by roadsides become the center of various infectious swarms of bees and other insects, which act as carriers of diseases like cholera, as well as stomach upsets, nausea, vomiting, etc.

The harmful toxin which comes from burning plastics causes carcinogenic diseases. Waste from chemical plants and nuclear plants mostly contains cadmium, arsenic, chromium nickel, dioxins, and PAHs, which are highly carcinogenic and can even create mutations in the genes and body. The central nervous system of the human body is also affected by these types of substances produced from waste. Pollutants like SO_2 and PM_{10} affect morbidity and mortality rates depending on the duration of exposure and mostly affect elderly people. Different types of chemical compounds like organo-chlorines and dioxins are lipophilic in nature and get accumulated inside the fatty tissues. This causes reproductive and endocrine problems. Studies show that low birth rates and neonatal deaths are commonly found near areas where large hazardous wastes have been dumped.

Waste and waste disposal create several other health issues, such as diseases in the respiratory system; irritation of exposed body parts like skin, nose, eyes, and ears; frequent headaches; gastric problems; stomach ulcers; fatigue; and allergies. It has been suggested that evaluation of a relationship between these symptoms is complicated and confounded by stress, public perception of risk, odors and nuisance related to the site, and recall bias. For example, a survey found that residents who indicated they were worried about pollution reported more symptoms than those who were not worried, both in the exposed and control areas.

2.2.1.3 Societal Impact

Waste affects the scenic beauty of a place. Peace and sanity of mind can also be disturbed. Studies show that people living close to waste disposal areas are psychological affected due to persistent bad odors spreading in the atmosphere. Town planning and city development authorities must consider waste disposal centers to be far away from townships. Municipal corporations are greatly responsible for the task of proper waste disposal, periodic garbage cleaning, and cleaning of drains and sewage systems. Even dust bins inside the city area must be regularly cleaned, and the garbage dumped at disposal yard must be regularly burnt to prevent any contamination or infectious diseases from spreading. Care should be taken that each individual uses proper dust bins and drainage system for disposing their household or office waste. This waste, if dumped openly outside, does not give a clean look and creates a lot of problems.

2.2.2 Hazardous Material in Waste

Landfill sites with waste disposal facilities emit different gases like carbon dioxide, methane, mercury vapor, hydrogen sulfide at low rates, and even 0.5% of volatile organic compounds. The pollutants present are mostly defined based on various parameters like level of toxicity, explosively, bioaccumulation, mobility, and environmental persistence. Not only this, even landfill sites consist of other materials like metals, polychlorinated biphenyls, pesticides, dioxins, asbestos, and pathogens. As per previous investigations, incineration of waste produces pollutants by burning sewage, and municipal and chemical waste.

Composting, sewage treatment, and landfills are also a source of hazards, containing microbial pathogens. Dust and the production of particulate matter are produced in landfill, incineration and composting processes, and by road traffic involved in all waste management types. Less easily quantifiable hazards, which might nevertheless impact on the population near a waste disposal site, include odors, litter, noise, heavy traffic, flies, and birds.

2.3 Waste Prevention

Waste reduction can be done at three stages of the product production: preliminary, intermediate, and final. Considering source waste production reduces the quantity and toxicity of waste before it undergoes other processes like composing, recycling, recovery of energy, and going to landfill. Waste prevention can be defined as the process of reducing the amount of materials used to create a product and increasing the efficiency with which the products, once created, can be used. There are different categories of waste prevention, for example quantitative and qualitative. When the hazardous content of the waste is removed, it greatly benefits humans and environment. Such a type of waste prevention method is quantitative, whereas just dumping huge amounts of waste to landfill is qualitative waste prevention.

2.3.1 Various Strategies for Waste Prevention

A study shows that there are three types of waste prevention strategies, as shown in Figure 2.2.

2.3.1.1 Informational Strategies

The most important concept of informational strategies is making a revolutionary change in people's minds and behaviors. Every human being lives their life and deals with their surroundings in their own way. By conducting several awareness programs, public awareness can be created that would people's mindset and behavior, help them to adapt to new changes, and make them more sensitive to the effects of their behavior on society. Some of the awareness programs that could be undertaken as informational strategies are:

- Regular campaigns in public places
- Informing the public about various waste prevention techniques
- Eco-labeling
- Societal training
- Incentives and rewards for a cleaner society
- Products with proper information for waste prevention
- Information about waste disposal

2.3.1.2 Promotional Strategies

This strategy deals with financial and logistic support for carrying out various waste prevention activities. Some of the activities are as follows:

- Voluntary agreement support system
- Facilities for eco-design and financial support

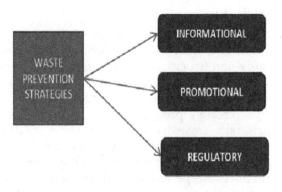

FIGURE 2.2
Three different strategies of waste prevention.

- Promotional activities on environmental management systems
- Infrastructure reuse, repair support systems
- Special incentives for clean consumption
- Emphasis on research and development activities

2.3.1.3 Regulatory Strategies

Regulatory strategies aim at enforcing various limitations on the production of waste, environmental obligation expansion, and establishment of environmental criteria on different public contracts. The various norms of this type of strategy include:

- Reform and changes in the responsibility criteria of the producer
- Procurement policies with involvement of green technology by the producer
- Taxes on waste generation

2.3.2 Waste Prevention Methods

There are various waste prevention methods possible, for example a) prevention, b) minimization, c) reuse, d) recycling, e) recovery, and f) disposal. Among these various methods, the first three methods are the most preferred, with disposal as the least preferred method, as seen in Figure 2.3.

2.3.2.1 Waste Prevention

Waste management always aims at the most fundamental aspect i.e., waste generation. If waste generation can be reduced, then automatically waste can be prevented. There are different techniques which can be adopted in the manufacturing sector, post-use, to reduce waste as well as pollution. Some of the common methods and strategies adopted in industries include adoption of environmentally friendly methods in working areas, use of less harmful and least hazardous materials, establishment of fire and leakage detection systems, improved quality of material storage, reduction in reactivity by incorporating chemical neutralization techniques, and different water saving methods. By adopting these all methods, waste can be prevented at some level at the initial and later phases.

2.3.2.2 Waste Minimization

It has been seen that in many manufacturing industries, such as operation, maintenance, or product delivery, waste cannot be totally eliminated from the system. This waste becomes a part of the system. Thus, in order to get rid of this waste, minimization techniques can be adopted, which will

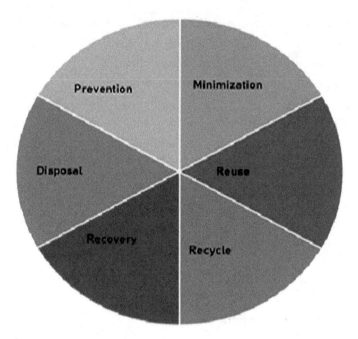

FIGURE 2.3
This shows various waste preventions methods.

decrease the total amount of waste released into the environment and, as a consequence, reduce pollution. Minimization of waste generation, or in other words source reduction, can be collectively utilized through the adoption of strategies of modern design, and the fabrication of products, goods, and services that minimize the quantity of waste generated and decrease the level of toxicity. Factories and large industries have adopted an alternative way of reusing materials, using substitute and less hazardous materials, using modern equipment, efficient machinery, and adopting modern procedures. In the process of waste minimization, a lower usage of resources would be greatly beneficial to the environment. In addition, this minimization of resources reduces the related costs.

Nowadays, modern packing systems have been developed that significantly reduce material use, save time, have better distribution systems, decrease fuel consumption, as well as reduce emissions to the atmosphere. Even building materials can be designed with special materials that reduce the overall mass and weight for a particular construction. In this way, unnecessary and excess materials can be reduced, and waste can be minimized.

2.3.2.3 Reuse

Reuse of products or materials is the phenomenon of using the material again and again until it is totally exhausted and has no benefits. There are various products which have been reused, and utility has increased for water in hydro power plants that can be used for electricity production as well as irrigation and other domestic activities. In addition, tall buildings and mansions can be made from different types of construction materials that can be reused, such as concrete, asphalt, reinforced steel, masonry, and bamboo poles. It has been seen that not only waste products can be reused but even by-products by refining and regenerating them. Some examples of such use include metal finishing processes where copper and nickel are recovered and using solvent extraction processes to recover oils, fats, and plasticizers. In the latter case, activated charcoal, clay, and sand are used as a filtering medium. Further, another method is the spray-roasting technique for acid recovery. Even using several bio-mass processes, food-based oils can be recovered and can be used as bio-diesel. Another example is waste from plants and trees that can be used as chips.

2.3.2.4 Recycling

Recycling is one of the methods of waste management which refers to recovery of different types of materials like paper, glass, plastics, wood, and metals from the waste section. These materials can be further processed, fabricated, and converted to a new finer product. Recycling of materials greatly reduces the use of original raw materials used for a particular product. This in turn reduces overall cost to the company in procuring new raw materials. Even different waste materials can be recycled, and with small or other changes can be made into a new form that then goes on to be used in other processes.

In addition, recycling has opened up advantages in various fields such as conservation of natural resources, reduction in energy consumption and environmental emissions, thereby reducing overall energy consumption and greenhouse gas emissions. Reduction of greenhouse gas emissions affects the global climate. Temperatures need to be at healthy levels for the survival of living organisms. Recycling can even create economic growth by creating several job options to carry out this work.

2.3.2.5 Recovery

Another strategy of waste prevention is recovery. Very often, many waste materials after reuse and recycling processes undergo different treatments to recover some of the useful products or by-products that can be used in various operations. Biological treatment is one such method of recovery of materials from waste. There are different methods of bio-degradation like aerobic composting, bio-treatment using mechanical methods, and anaerobic digestion. Usable compost is made by degradation of waste i.e., by separating the organic part from the inorganic material and allowing aerobic composting.

Animal manure, waste from yards, and food are known as organic waste. This waste contains bacteria that are degrading in nature and can convert this waste to compost, which can then be utilized as fertilizer. Even aerobic composting is a method of recovery of waste to create another product. This is done by piling up organic wastes in open areas or sometimes close together so that gas can be collected. In this process, wood chips are added to the waste material so that aerobic degradation of organic materials can be better. Finally during curing process where pathogens are destroyed, the final material is matured and stabilized. Along with the final product, carbon dioxide and water are also produced.

Compost materials are mostly used for improvement and remediation of soil and ground water. These processes are labor intensive. The quality greatly depends on the quality control of the composting process adopted. Sometimes, if the process is not properly conducted with good quality control, large amounts of carbon dioxide are released into the atmosphere. Similarly, the anaerobic degradation process releases methane, carbon dioxide, and bio-solids. Even biogas has wide applications, for example it can be used for heating and electricity, and the residues can be used as natural fertilizers. The most important advantage of anaerobic digestion is biogas collection. In this process, degradation of waste is much faster than landfill disposal.

Incineration is a process of waste combustion at a very temperature to produce electrical energy. This process creates ash as a by-product and reduces various kinds of hazardous waste like chlorinated hydrocarbons, oils, solvents, and pesticides.

2.3.2.6 Disposal

Under waste management schemes, disposal is one of the methods of waste prevention. Thus landfills are methods of waste disposal. Landfills are designed to receive waste of hazardous types as per RCRA subtitle C regulations and also municipal solid waste as per subtitle D regulations. A landfill can consist of processes regarding the collection of leachate gas collection systems, and even monitoring of ground water. Landfills cannot be directly managed by any individual; this requires special permission. Waste within the landfill sites becomes degraded anaerobically. The output of the degradation process is biogas, which can be collected and used for other utility purposes. Co-generation systems can be combined to produce heat and electricity. Landfills can be transformed by land recycling methods to create wonderful playgrounds, recreational parks, and golf courses.

2.3.3 Analysis of Waste and Waste Prevention

For understanding the effect of waste and its prevention methods on society, a sample of 100 people each was taken for two different areas at

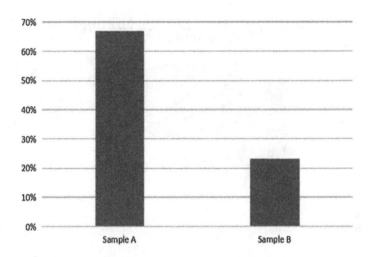

FIGURE 2.4
Two samples of data showing the health effects.

Bhubaneswar, Odisha. Sample A consists of 100 people residing near waste landfills and sample B consists of 100 people far away from waste landfills. The sample for the test was taken from a group of people residing in the apartments and buildings near these two defined areas. A close health study was conducted for one year, and from the data collected from the health reports of the nearest hospitals or dispensaries, it is clearly shown that people staying in the close vicinity of the waste disposal area are affected by more types of diseases than those living further away. Even changes in the psychological behavior are observed, and the atmosphere remains polluted. The worst sufferers are the waste management workers who dwell in these places. From the analysis it is observed that around 67% of health cases are observed for sample A and 23% from sample B. The results are presented in Figure 2.4.

2.4 Case Study: Three Rs of Waste Prevention

Among the different waste prevention methods discussed in this chapter, the three Rs are the typical three methods which every organization should follow, as seen in Figure 2.5:

1. R – Reduce
2. R – Reuse
3. R – Recycle

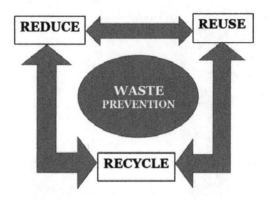

FIGURE 2.5
The three Rs concept of waste prevention.

We consider a case study at the Kalinga Institute of Industrial Technology School of Electrical Engineering, Bhubaneshwar, Odisha, where this three R concept is implemented and is found to have a substantial benefit to the school in terms of both money and environment concerns. A period of two years, i.e. 2018 and 2019, was considered for analysis of the effect of adopting these three R waste prevention methods. In 2018 waste prevention methods had not been adopted, whereas in 2019 waste prevention methods were adopted.

From one of the waste prevention methods, "REDUCE", some of the measures taken are, for example:

- Unnecessary Xerox and office notices are prevented and e-notices are preferred
- For advertisements, digital marketing is considered
- Publication of regular college articles in online system and e-magazines

Similarly, from the other method, "REUSE", the various methods adopted are for example:

- Reuse of Xerox papers
- Reuse of project components in making other projects
- Reuse of banners and posters by Krutovites society
- Reuse of marker pens for writing on the board

For the last R of waste prevention method, "RECYCLE", the various measures taken are for example:

- Recycling of student answer sheets, assignments, project reports, seminar reports
- Recycling of cardboard/dusters

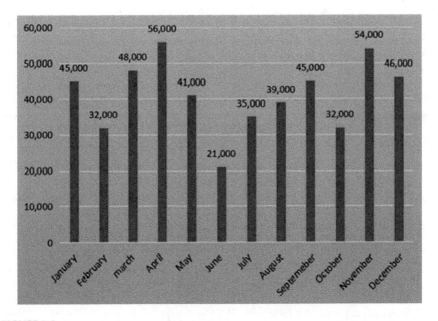

FIGURE 2.6
Monthly savings for 2019.

After using all these methods, a financial analysis was done to find out what the savings of the department by adopting these three Rs waste prevention methods were. This is presented in Figure 2.6. The Figure shows the savings of the department each month by using three Rs waste prevention method.

2.5 Conclusion

Waste generation is unavoidable and exists with the growth of industrialization. As society grows and contemporary society uses up-to-date equipment and services, waste grows simultaneously. In this study we have seen the impact of waste on various sectors like health, environment, and the economy. Better waste management policies adopted by organizations can help to get rid of some of the hazards generated with waste disposal. Education and awareness have to be implemented widely in society to make people better able to deal with waste hazards and to adopt practices and methods to reduce waste generation. Every organization should make adoption of waste management policies compulsory. All methods, like recycle, reuse, and reduce, can be executed well in every sector. Adoption of the seven steps of methodical assessment for energy management can really impact waste prevention. The steps are the first and most important part of identification of waste management options, defining functional units and setting limits for boundaries, evaluating environmental performance, evaluating economic performance, choosing optimal scenarios, and finally analyzing the sensitive effect.

References

1. Wilts, Henning. "National waste prevention programs: Indicators on progress and barriers." *Waste Management & Research* 30.9_suppl (2012): 29–35.
2. Bartl, Andreas. "Moving from recycling to waste prevention: A review of barriers and enables." *Waste Management & Research* 32.9_suppl (2014): 3–18.
3. Van Ewijk, S., and J. A. Stegemann. "Limitations of the waste hierarchy for achieving absolute reductions in material throughput." *Journal of Cleaner Production* 132 (2016): 122–128.
4. Puig-Ventosa, Ignasi, Marta Jofra-Sora, and Jaume Freire-González. "Prevention of waste from unsolicited mail in households: Measuring the effect of anti-advertising stickers in Barcelona." *Journal of Material Cycles and Waste Management* 17.3 (2015): 496–503.
5. Williams, I. D., F. Schneider, and F. Syversen. "The 'food waste challenge' can be solved." *Waste Management* 41 (2015): 1–2.
6. Sharp, Veronica, Sara Giorgi, and David C. Wilson. "Methods to monitor and evaluate household waste prevention." *Waste Management & Research* 28.3 (2010): 269–280.
7. Beigl, Peter, Sandra Lebersorger, and Stefan Salhofer. "Modelling municipal solid waste generation: A review." *Waste Management* 28.1 (2008): 200–214.
8. Cox, Jayne, et al. "Consumer understanding of product lifetimes." *Resources, Conservation and Recycling* 79 (2013): 21–29.
9. Sharp, Veronica, Sara Giorgi, and David C. Wilson. "Delivery and impact of household waste prevention intervention campaigns (at the local level)." *Waste Management & Research* 28.3 (2010): 256–268.
10. Salhofer, Stefan, et al. "Potentials for the prevention of municipal solid waste." *Waste Management* 28.2 (2008): 245–259.
11. Bernstad Saraiva Schott, Anna, et al. "Potentials for food waste minimization and effects on potential biogas production through anaerobic digestion." *Waste Management & Research* 31.8 (2013): 811–819.
12. Gentil, Emmanuel C., Daniele Gallo, and Thomas H. Christensen. "Environmental evaluation of municipal waste prevention." *Waste Management* 31.12 (2011): 2371–2379.
13. Curran, Anthony, and Ian D. Williams. "The role of furniture and appliance re-use organisations in England and Wales." *Resources, Conservation and Recycling* 54.10 (2010): 692–703.
14. Cox, Jayne, et al. "Household waste prevention: A review of evidence." *Waste Management & Research* 28.3 (2010): 193–219.
15. Bulkeley, Harriet, and Nicky Gregson. "Crossing the threshold: Municipal waste policy and household waste generation." *Environment and planning A* 41.4 (2009): 929–945.

16. Kissling, Ramon, et al. "Definition of generic re-use operating models for electrical and electronic equipment." *Resources, Conservation and Recycling* 65 (2012): 85–99.
17. James, K. "A methodology for quantifying the environmental and economic impacts of reuse." *Final report. WRAP* (2011).
18. Truttmann, Nina, and Helmut Rechberger. "Contribution to resource conservation by reuse of electrical and electronic household appliances." *Resources, Conservation and Recycling* 48.3 (2006): 249–262.
19. Gelbmann, Ulrike, and Barbara Hammerl. "Integrative re-use systems as innovative business models for devising sustainable product–service-systems." *Journal of Cleaner Production* 97 (2015): 50–60.
20. Christis, Maarten, et al. "Value in sustainable materials management strategies for open economies case of Flanders (Belgium)." *Resources, Conservation and Recycling* 103 (2015): 110–124.

3

Smart Waste Bin Using AI, Big Data Analytics and IoT

Nivedita Das, Jyotiprakash Panigrahi,
Chandrima Roy, and Biswaranjan Acharya

CONTENTS

3.1 Introduction

Waste material that is discarded by people, typically due to a perceived lack of utility, is garbage, waste, rubbish, or refuse. The definition does not necessarily include products of bodily waste, solely liquid or gaseous waste, or products of hazardous waste. Garbage is typically sorted into kinds of material appropriate for various methods of disposal and is graded. Burying waste also causes both air and water contamination, and a growing amount of valuable fossil fuels are used by merely shipping it to various sites, resulting in more pollution and other problems [1]. The average plastic garbage bag, buried in a landfill, needs 1,000 years to decay, giving off contaminants as it does. At present, the collection and maintenance of the municipal waste management system are essential activities in order to keep high hygienic standards. The use of the conventional method of waste collection results in an ineffective system and wasting time and resources [2].

3.1.1 AI in Smart Waste Management

The existing waste management systems are incapable of effectively dealing with the massive amounts of garbage produced every day. By moving to AI [3] for smart recycling and waste management, garbage sorting and disposal processes will be automated, resulting in more sustainable recycling practices.

Figure 3.1 shows how artificial intelligence helps in waste management starts with intelligent garbage bins. Waste management companies use Internet of Things (IoT) sensors to keep track of how complete trash cans are in the region. Municipalities may use this information to improve waste collection paths, times, and frequencies.

Figure 3.2 represents various uses of AI in waste management. Various concepts about this are described as follows:

1. With the advent of RFID tags, waste sorting systems have changed dramatically. Songdo, a city in South Korea, employs RFID tags to sort trash into different categories. The tags are then read by a

FIGURE 3.1
How does AI help in waste management?

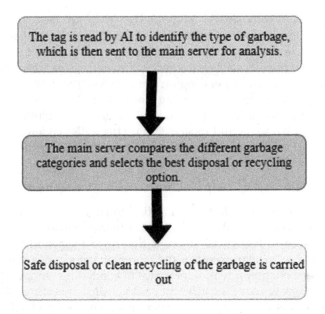

The tag is read by AI to identify the type of garbage, which is then sent to the main server for analysis.

The main server compares the different garbage categories and selects the best disposal or recycling option.

Safe disposal or clean recycling of the garbage is carried out

FIGURE 3.2
Uses of AI in waste management.

pneumatic waste disposal system. As a result, the primary server, which stores all of this information, determines the best way to dispose of all of the waste generated.

2. Another revolutionary concept in waste management is the intelligent trash-can, which is equipped with AI programs and IoT sensors. The sensors on these trashcans measure the waste levels of the garbage that is thrown in and send this information to the main disposal system for processing through intermediate servers. The data are organized by garbage type, quantity, and waste disposal method. This system as a whole should improve its effectiveness over time by analyzing historical records.

3. Waste sorting robots have begun to be used in landfill sites. Traditional waste sorting methods are being phased out in favor of automated intelligent machines. The robots can sort tons of garbage in a day, thanks to their multitasking abilities. These robots can easily differentiate between tin foil and paper thanks to their computer vision programs. Such large-scale systems have enormous potential for use in a wide range of industries.

3.1.2 AI in Smart Recycling

Two SFU Mechatronics Systems Engineering engineers created an AI-powered smart recycling machine that will revolutionize waste management and enterprises. While sorting and sending garbage, the intelligent bins should reason for themselves. All that is required is for the trash to be deposited in the appropriate waste bin. Before determining what to do with the garbage, the bin uses its sensors to inspect and equate the trash retrieved in previous trash records. Depending on the decision, the garbage is directed to the appropriate disposal system, such as a landfill or a recycling facility. We can expect a substantial reduction in waste generated globally if we find better ways to dispose of and recycle trash. This will go a long way toward preserving the environment for a more prosperous and sustainable future.

3.1.3 Big Data Analytics Helps in Smart Waste Management

The introduction of big data has simplified the entire recycling and waste management process. big data has affected many industrial and scientific fields [4], as seen in Figure 3.3, in ways that favor the world [5]. It is used daily, for instance, to help scientists and land managers gain a deeper understanding of our changing world and ways of combating climate change [6]. One way to improve recycling efficiency is to create a recycling robot, which makes the process of sorting solid waste much cheaper and safer. The big data analytics used for different societal applications are defined in the literature [7], which interested readers can read.

It is through gathering information related to the shapes, textures, and even brand logos of the content [8] it processes that the robot works. In total, almost 60 cartons of recyclables per minute can be sorted by a robot! One way big data is making recycling more effective is through the invention of

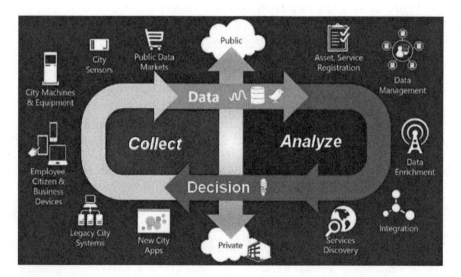

FIGURE 3.3
Concepts of big data.

a recycling robot, which makes processing solid waste much cheaper and safer [9]. The content is sorted by the robot using data related to its shapes, textures, and even brand logos [10].

Big data [11] is frequently used to aid route planning in order to obtain better estimates of how much waste is made, as shown in Figure 3.4. One waste management company in Manchester, for example, uses big data to figure out which neighborhoods generate the most waste and then targets those neighborhoods for better recycling education. Big data may also be used to help large enterprises define areas in which they produce waste in order to establish strategies for waste reduction [12]. By producing goods that are less costly or environmentally damaging, many major corporations are attempting to become more sustainable [13].

Decreasing the amount of manufacturing waste created during product production is one way businesses can become more sustainable. Here, big data can aid in increasing productivity and cataloging [14] ways to minimize packaging materials without hazardous items. This will help companies save money in the long run by reducing their resource use significantly. Figure 3.5 represents the various big data applications in smart cities.

3.1.4 Internet of Things (IoT) Helps in Smart Waste Management

To put it simply, IoT is a network consisting of hundreds of devices that can connect with each other. The IoT's AI and ML capabilities allow the network

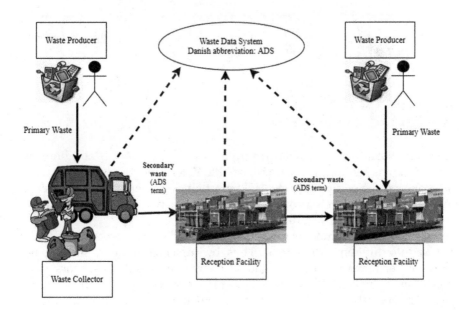

FIGURE 3.4
Garbage collection in big data environments.

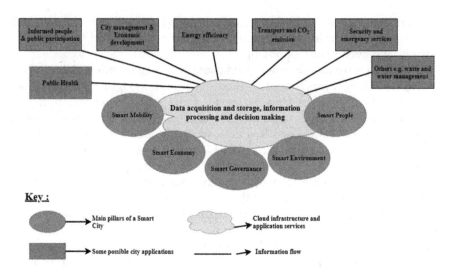

FIGURE 3.5
Big data applications to smart cities.

FIGURE 3.6
How does IoT help in waste management?

to process the information obtained from the linked gadgets. In addition, this information is sent to the user or is used to determine the next action, such as changing the equipment, etc. [15] Smart cities, as depicted in Figure 3.6, are cities that work to improve the lives of their citizens through the use of digital technology and big data.

It is difficult for almost every country to handle the increasing influx of people into different urban areas. Thus, by implementing data sharing and analytics, artificial intelligence, and, of course, thousands of sensors, cities are becoming smarter. New innovations to save operating costs and optimize the value of existing assets are being implemented by local corporations.

This paper will first introduce the concepts of artificial intelligence, big data, and IoT in Section 1, which will help in Smart Waste Management.

Section 3.2 is about previous studies. Section 3.3 is about the methodology of the proposed architecture and Section 3.4 shows the implementation and how actually it works. In Section 3.5, we analyzed the result of this technique, and Section 3.6 is about the conclusion of the chapter and future scope.

3.2 Related Work

Waste management has been the subject of extensive study all around the world. The majority of studies have concentrated on waste generation, distribution, and recycling manually. People have recently begun to employ automation to efficiently track and capture waste. Several IoT-based smart technologies have been proposed to solve various issues with smart city waste management systems. The most important problem for the smart city is solid waste management. To address these problems, especially solid waste management, the researchers have used a variety of methods and techniques. Power, weight, temperature, humidity, and chemical sensors are used in the monitoring and processing of solid waste [16].

The authors proposed a municipal solid waste management platform for recycling collection information in [17], with the help of IT technology. A model for waste management, transportation, recycling, and disposal was built in this report. The findings of this study show that the built framework aids municipal authorities by making use of the data produced at each level of waste monitoring and processing. Finally, the scheme met its goal by creating an adaptive waste disposal system for collection materials. Mahajan et al. have suggested a waste bin control method based on Zig-Bees [18]. The sensors in the garbage bins detect how much trash is in the container, and the information is sent to the garbage collection truck driver through a fast messaging service. In Bhor's proposed system [19], the amount of trash in the bins is detected using sensor systems and communicated to the authorized control room via GSM unit. A GUI is also being created to track the necessary details related to waste bins at different locations. Garbage collecting would be more efficient as a result of this. IoT-based smart garbage and waste disposal bins have been suggested by S.S. Navghanya. The waste bins are attached to a microcontroller using a wireless IR device in this system. Via Wi-Fi, the webpage provides updates on the state of the garbage. The IR sensor is used to provide information about the various amounts of garbage in the container. When the threshold level is crossed, the weight sensor is engaged. Instead of using costly smart bins, they propose to use a sensor-based solution that is less costly. They often use three IR sensors to show each amount in the dustbin, as well as a Wi-Fi router to access real-time bin status [20].

TABLE 3.1

Studies of Previous work

Objective	Components Used	Type of Segregation	Techniques Used
Automated waste segregator [22]	Flaps, motors, Capacitive plates, Inductance coil, bins, IR sensor	Wet waste, dry waste, and metal waste	Waste is detected by an infrared proximity sensor in the device. Metal detector sensor distinguishes between wet and dry waste using a capacitive sensing module.
The research on the relationship between waste classification [23]	Jumper cables, Arduino, ultrasonic tracker, Servo motors, 6V power supply	Environmentally friendly	It will collect household waste such as paper, plastics, and cardboard from a single location and deposit it in the classifier platform.
Designing waste disposal schemes and packaging waste [24]	flaps, motors, Arduino, IR sensor	Wet waste, dry waste	Within the waste collection issue, the concept of dynamic routes was discussed. Although real-time data were used in this study, it concentrated on a simplified version of the problem in which an optimization model is solved every day with just 68 waste bins
Waste management of smart city [25]	Arduino, IR sensor	Wet waste, dry waste	The ultrasonic sensor detects the amount of waste present and compares it to a set of standards. The sensor sends a message to the Arduino, which is in control of the system's overall communication, if the current amount of waste reaches the threshold limit

Ranchan Mahajan has suggested an IoT-based waste bin monitoring system. The sensors are mounted in public garbage bins, and the ARM7 controller is informed of the threshold level indication. The controller can inform the garbage collection truck driver which garbage bin is fully full and requires immediate attention. The ARM7 will show a message sent using GSM technology [21]. Some more research on this is described in Table 3.1.

3.3 Methodology

The proposed model includes the inspection and disposal of solid waste from wet and dry waste bins located in a locality. The proposed system's

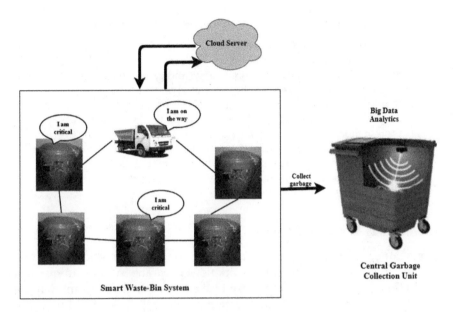

FIGURE 3.7
Proposed architecture of a smart waste bin.

architecture is shown in Figure 3.7. The proposed framework is applied in the following stages:

1. **Construct a hardware module to synchronize sensor data of the waste bins.**

   Garbage bins are mounted at each place, each with ultrasonic sensors that signify the amount of garbage in the bin. The waste bin status is sent to a central cloud server using an IR level sensor and humidity sensor.

2. **Software and hardware module development on a cloud platform.**

   This module will receive and display real-time status updates from all of the garbage bins.

3. **Development of a map-based application that shows the current location and condition of trash cans.**

   The app will display garbage bin status in real time, which will be used by garbage collection truck drivers to determine when bins are full or likely to be full.

4. **The shortest path between the garbage truck and fully loaded garbage bins is dynamically generated and displayed.**

   The final stage determines the shortest path between the garbage truck and the garbage bins. A central cloud server takes care of route optimization.

Algorithm 1: Smart Waste Management Algorithm

Step 1: Initialize deployed waste bins in different location
Step 2: Initialize web socket network bins
Step 3: Gather all level sensor data, humidity sensor data, and thermal sensor data from the integrated center
Step 4: Synchronize information every hour on a cloud server
Step 5: Transmit the collected data over the internet to the servers
Step 6: Store and process the information on the server after the prediction model predicts the critical percentage and if critical then schedule for pick-up, otherwise wait for critical
Step 7: Using the Dijkstra algorithm, find the shortest path and determine the critical waste bin

3.3.1 Support Vector Machine

The machine learning algorithm known as "Support Vector Machine (SVM)" was used to carry out the research work flowchart. The SVM technique was used in this study. SVM is a supervised learning method algorithm that can be applied to regression and classification problems [26]. But SVM has been mostly used in classification problems. SVM is a linear classifier. It predicts results with both training set and test set. It performs classification by finding the hyper plane that differentiates the two classes perfectly. How the SVM classifies the hyper plane is shown in Figure 3.8. It actually classifies the data set before tuning.

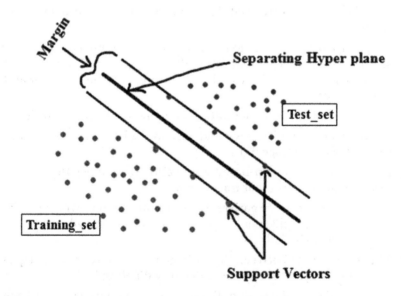

FIGURE 3.8
Two lines separating one hyper plane with the help of training and test set. SVM with separating hyper plane.

Algorithm 2: Support Vector Machine (SVM) Classification Algorithm

Step 1: data = Load raw data()
Step 2: trainset = data*0.8
       testset = data*0.2
       trainset Lable = trainset[Last Column]
Step 3: SVM.initialize()
       SVM.fit(trainset,trainset Lable)
       SVM.save("svm.pkl")
Step 4: SVM.load model()
       SVM.predict(tesetset)

3.3.2 Arduino

Arduino is an open-source electronics platform with simple hardware and software that is low-cost. Arduino boards can take inputs like light from a sensor, a finger on a button, or a tweet and transform them to outputs like turning on an LED, activating a motor, or publishing something online. Figure 3.9 depicts the Arduino board that was used.

3.3.3 GPS Module

The Global Positioning System, or GPS, is a satellite-based global navigation system that provides synchronized direction, velocity, and time. Everywhere

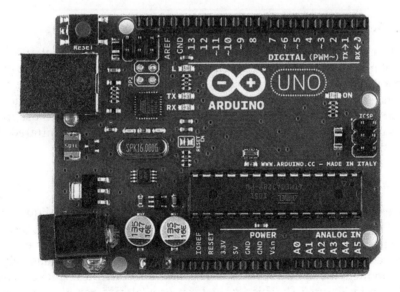

FIGURE 3.9
The Arduino Uno is an open-source microcontroller board designed by Arduino.cc and based on the Microchip ATmega328P microprocessor.

FIGURE 3.10
The Global Positioning System (GPS) was created to allow military and civilian users to accurately determine geographical locations.

you look, GPS is present. GPS devices are used in cars, smartphones, and watches. The longitude and latitude of the smart bin are calculated using GPS here. The GPS is shown in Figure 3.10.

3.3.4 Sensor

A sensor is a device that senses sensory data in its environment and converts it into data that humans or machines can understand. The majority of sensors are electronic (data are converted to electronic data), but some, such as a glass thermometer that displays visual data [27], are simpler. Different sensors are used for different purposes. Here in this chapter, we use three types of sensors. These are level sensor, humidity sensor, and infrared (IR) thermal MLX 20 sensor.

3.3.4.1 Level Sensor

A level sensor is a system that monitors, maintains, and measures liquid (and sometimes solid) levels [28]. Once the liquid level has been detected, the sensor converts the perceived data into an electric signal. There are seven main level sensors. These are capacitance, ultrasonic, microwave/radar, vibrating, conductivity, optical level switches, and float switches. A level sensor is shown in Figure 3.11. This IR level sensor is always generic; no specific model is used. It checks the level of the smart bin, meaning it checks the status of the bin. The status is semi-critical, critical, and not-critical [29].

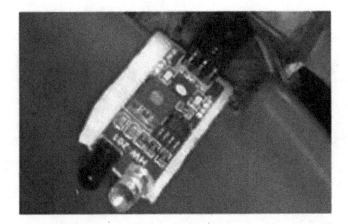

FIGURE 3.11
IR Level Sensor for Arduino with IR LED.

3.3.4.2 Humidity Sensor

A humidity sensor (also known as a hygrometer) is an electronic system that detects humidity in its surroundings and transforms the results into an electrical signal [30]. To determine relative humidity, the maximum amount of humidity for air at the same temperature is compared to the live humidity reading at the same temperature. A humidity sensor detects monitors and records moisture as well as air temperature. The ratio of moisture in the air to the maximum amount of moisture at a given air temperature is called relative humidity [31]. When looking for warmth, relative humidity becomes a key consideration. A humidity sensor (also known as a hygrometer) is shown in Figure 3.12. This sensor measures the percentage of humidity inside or outside of the bin (very near area).

3.3.4.3 Infrared (IR) Thermal Sensor

These sensors concentrate infrared energy from an object onto one or more photodetectors [32]. These photodetectors translate the energy into an electrical signal equivalent to the
target's infrared energy. Different types of infrared (IR) thermal sensors are present. Some of them are very popular. These are Infrared (IR) Thermal MLX 20 Sensor, Infrared (IR) Thermal MLX 40 Sensor, Infrared (IR) Thermal MLX 80 Sensor, and Infrared (IR) Thermal MLX 120 Sensor. Here in this chapter, we used Infrared (IR) Thermal MLX 20 Sensor as shown in Figure 3.13. This sensor measures the temperature [33].

3.3.5 Functions of Proposed System

In our experiment we follow these methods or functions as shown in Figure 3.14. The main functions are given below in Figure 3.14.

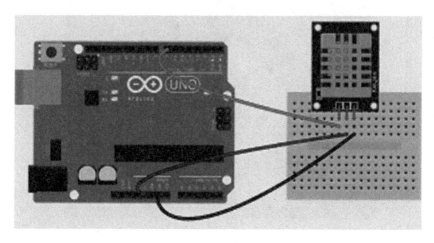

FIGURE 3.12
IR Level Sensor for Arduino with IR LED.

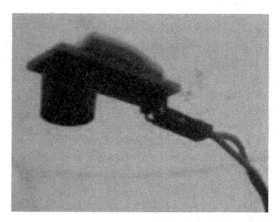

FIGURE 3.13
Infrared (IR) Thermal MLX 20 Sensor.

a) **Dataset**

   When we start generating data using all sensors with GPS, the produced data are a combination of ordinary values and floating-point values. To get a dataset (as shown in Figure 3.18) we need to do pre-processing.

b) **Pre-processing**

   This involves data being pre-processed in order to allow it to be further examined, or primary processing. Pre-processing steps could include things like extracting data from a larger collection, filtering it for different purposes, and merging datasets. In this process the

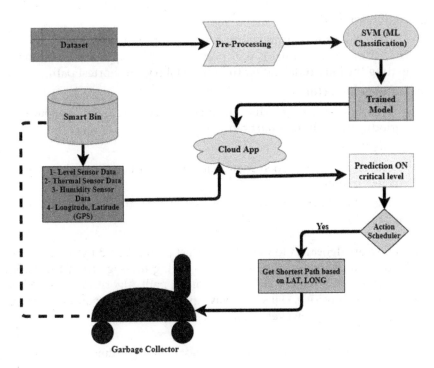

FIGURE 3.14
Functions of the proposed system.

generated data that show ordinary value with string and floating point are converted to normal form (as shown in Figure 3.18).

c) **Machine learning classification (ML) technique**

Here a machine learning classification technique known as a Support Vector Machine (SVM) is used [34]. The produced dataset is non-linear, so we cannot use any linear algorithm. So, we use the SVM algorithm. The behavior of SVM has always given the best performance with a smaller number of datasets. In our process, we have only 400 attributes of data.

d) **Trained model**

After implementing SVM and tuning, a model is generated. Here we get an efficiency of trained model.

e) **Cloud app**

After getting trained model data with accuracy, it goes into a cloud app. All sensor data with GPS data also goes into a cloud app.

f) **Prediction**

The process of predicting what will occur in the future is known as predictive. It will check the critical level.

g) **Action scheduler**

Action scheduler is performed according to the predicted value, i.e., critical level status. It will prepare for pick-up operation after getting the longitude and latitude from GPS with shortest path.

h) **Garbage collector**

This is removing or discharging garbage and frees up spaces periodically from the smart bin.

i) **Smart Bin**

This is an automated system that collects garbage in a smart and efficient manner.

The flowchart for the proposed architecture is shown in Figure 3.15.

Here we use an algorithm to find the shortest path covered by the cleaning vehicle to clean the smart dustbin after a getting message from the cloud. The algorithm is known as "Dijkstra's shortest path Algorithm" [35]. When all edge costs are positive, Dijkstra always gives the shortest path. When negative edge costs exist, however, it can fail.

Algorithm 3: Dijkstra's Shortest Path Algorithm

Dijkstra(G, w, s)

Step 1: INITIALIZE SINGLE_SOURCE (G, s)
Step 2: s = Φ <- set of evaluated vertices
Step 3: Q = G.v <- minimum priority queue
Step 4: While Q ≠ Φ
Step 5: u = EXTRACT - MIN(Q)
Step 6: s = s ∪ {u}
Step 7: For each vertex v ∈ G. adj[u]
Step 8: RELAX(u, v, w)

The vehicle follows Dijkstra's algorithm and gets ready to perform (as shown in Figure 3.16(a)).

After receiving a message from the cloud, the vehicle begins to operate. After determining the shortest route, the vehicle travels from point A (start) to point E (destination) (as shown in Figure 3.16(b)).

For calculating the best route for a cleaning vehicle using GPS, find the final shortest path in which the garbage collector vehicle arrives faster (as shown in Figure 3.16(c)).

3.3.6 Case Study for Critical Percentage

Here we create real-time data sets to train the machine. To train the machine we need the critical percentage. Case studies are given in Table 3.2.

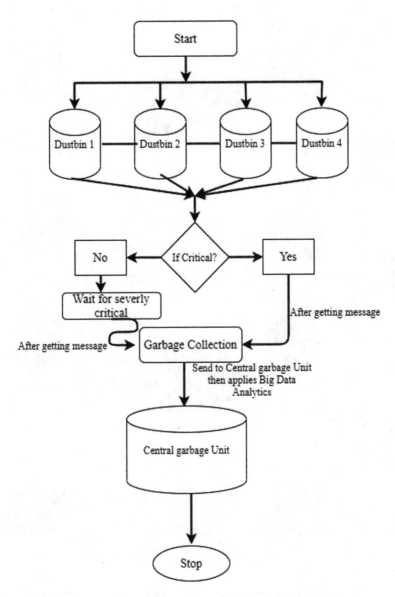

FIGURE 3.15
Flow-chart of proposed architecture.

3.4 Implementation

As a working model of the real world, the proposed framework is tested in practice. Here we take three pieces of data and produce a dataset which trains our whole system. According to critical percentage, the smart bin

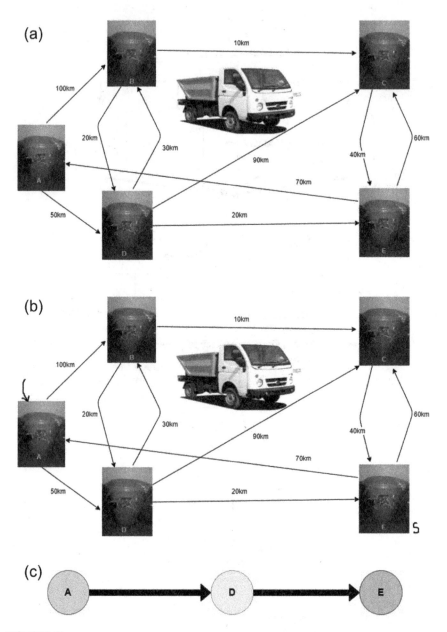

FIGURE 3.16
(a) Dijkstra's algorithm gets ready to perform. (b) Following Dijkstra's algorithm and starting to work after getting message (source-sestination(s)). (c) The vehicle arrives and follows this shortest path route.

TABLE 3.2

Case Study for Critical Percentage

Case	Level (0–100%)	IR Thermal Temp	Humidity	Critical Type
1	No	Low	High	Semi-critical
2	Less	High	High	Semi-critical (rotten and bad smell)
3	Full	Low	Low	Critical (immediate action required)
4	Less	High	Less	Semi-critical

(Note: If it is very hot and the level is not full at all, then no action is required. For example, generally at 12 noon it is very hot. The sensor is synchronized to the main board every hour.)

requires cleaning as soon as possible. When it takes data from these discussed three sensors (critical level, humidity level, and thermal signature), it is connected to the Arduino board. Then the Arduino board is connected to the GPS. Where through GPS, we show longitude and latitude of smart bins. This board connects to Wi-Fi. So, this system works where Wi-Fi is available.

Here we pulled data, and we send it to the application which has already been placed in the cloud. We placed the critical level percentage in the cloud. After getting numbers of data, we trained with the help of a machine learning model. After training, it will define whether the bins are actually critical or not. In every hour, a schedule takes place in an Arduino board. All data coming from all sensors are managed by the cloud. Generally, at 12 noon, heat level is high. Critical level is also high due to the heat. But we don't need to clean the dustbins. To avoid this type of situation we used artificial intelligence. Figures 3.17(a), (b), and (c) show the setup of the smart bin.

When the dustbin shows as critical it sends messages to the cloud. Then the cleaning vehicle gets a message from the cloud. It moves and follows Dijkstra's shortest path algorithm. These smart bins are always placed in a Wi-Fi area. They only work in smart cities. If we have a neural network, it will directly connect without the need for Wi-Fi.

3.5 Results Analysis

We produced the dataset shown in Figure 3.18 using this sensor data. But the generated dataset is in the form of alphanumeric and floating point. After pre-processing we get this dataset. After getting the dataset, we train our model using machine learning classification, i.e., support vector machine (SVM). We have 400 attributes of data, so SVM gives the best performance. Figure 3.18 represents the screenshot of ten datasets. After implementing

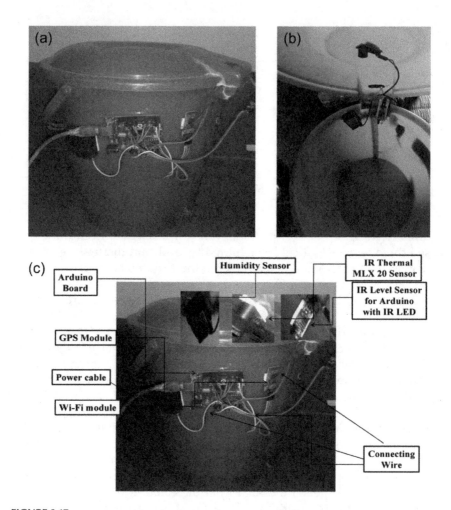

FIGURE 3.17
(a) Set up of outside of smart bin. (b) Set up of inside of smart bin. (c) Set up of whole smart bin with equipment's name.

SVM and tuning, a model is generated. The flow time efficiency of the trained model is 90%.

3.5.1 Dataset Description

In this case, temperature is a critical factor. We don't wait until the trash can is full. We decide whether or not to pick up the dustbin right away based on sensor data. If a dog's body is dumped in a dustbin, no one will use it because of the foul odor. It actually needed to be picked up right away. Otherwise, it creates an unfavorable environment. We installed a sensor inside the dustbin that indicates if it is not critical and is also not full. As a result, no action is

	A	B	C	D	E
1	**Humidity(%)**	**Temprature(°C)**	**Label**	**Class**	
2	30	30	Full	Semi-Critical	
3	63	30	Not-Full	Critical	
4	40	37	Not-Full	Semi-Critical	
5	81	33	Not-Full	Semi-Critical	
6	46	31	Full	Critical	
7	90	26	Not-Full	Critical	
8	77	35	Full	Critical	
9	58	50	Not-Full	Critical	
10	27	25	Not-Full	Not-Critical	
11	10	11	Not-Full	Not-Critical	
12					

FIGURE 3.18
Screenshot of real-time dataset generated before trained model (in .csv).

required. Any parameters are not hard coded in this case. It accurately predicted all of the parameters with generated values. Our goal is to answer the question, "How do we handle waste wisely?" In this data set, we describe the entire data generation process.

a) The first data represents that the humidity is 30%, the temperature is 30°C, and the level is full, indicating that the class is semi-critical, which means that only dry waste is present in the dustbin at this time. It is ready for pickup, but if it is not picked up soon, it is fine.

b) The humidity is 63%, the temperature is 30°C, and the level is not full, so the class is critical, indicating that some rotten things were recently placed in the dustbin. As a result, the humidity rises. Pick-up action is required as soon as possible.

c) The third data shows that the humidity is 40%, the temperature is 37°C, and the level is not full, indicating that the class is semi-critical, implying an unnecessary increase in temperature at 12 noon.

d) In the fourth data point, humidity is 81%, temperature is 33°C, and the level is not full, indicating that the class is semi-critical, implying that some liquid things like curry or water have been thrown away. So, it's better if the action is completed.

e) The fifth data point shows that the humidity is 46%, the temperature is 31°C, and the level is full, indicating that the class is critical and that the dustbin requires immediate attention.

f) In the sixth data point, the humidity is 90%, the temperature is 26°C, and the level is not full, indicating that the class is critical, indicating that the dustbin is full of watery things. It calls for immediate action.

g) The seventh data point shows that the humidity is 77%, the temperature is 35°C, and the level is full, indicating that the class is critical and that immediate action is required for the dustbin. The eighth data point takes humidity is 58%, the temperature is 50°C, and level is not full, so the class is critical, meaning the dustbin is full of rotten things. It requires immediate action.

h) The ninth sample point shows humidity at 27%, temperature at 25°C, and level is not full, indicating that the class is not critical and no action is required for the dustbin.

i) The tenth data point represents the humidity at 10%, the temperature at 11°C, and the level is not full, so the class is not critical, which means the dustbin does not require immediate action.

3.6 Conclusion and Future Work

When a smart-bin is introduced into our daily lives, it will be extremely beneficial in avoiding the creation of a bad odor or a bad scene. Here are some real-world examples. When someone throws dead animals or birds into a bin, this emits a foul odor and creates a disturbing scene. This is where the smart bin comes in handy. It sends a message to the cleaning vehicle when it is full. There's no need to wait for municipal sweeping crews.

Smart solid waste management programs include real-time intelligent bin status monitoring and rule-based decision algorithms. For sending solid waste data through a wireless sensor network, the monitoring program uses decision algorithms. The machine is composed of three components: a smart bin, a portal, and a control station. The basic idea is that when the waste bins reach the critical level, smart bins gather their status data and send it to a server via a coordinated intermediary. A series of server applications display the modified bin status in real time. The main contribution of the system is the development and implementation of an automatic bin status management system that employs a variety of novel rules-based decision algorithms. However, the device faces technological challenges such as gateway long-range connectivity technology, incorrect sensor data output, and a lack of GPS for position tracking, as well as citizen participation in the system for better interactions.

In the future this can be developed into edge-based distributed smart waste bins with the feature of recycling. When it comes to separation of materials, the recycling material will be recycled at another location. This will prevent environmental pollution as well as waste.

References

1. Panigrahi Jyotiprakash, Bhabani Shankar Prasad Mishra, and Satya Ranjan Dash. "Disease prediction on the basis of SNPs." In *Emerging Technologies in Data Mining and Information Security*. Springer, Singapore, 2019. 635–643.

2. Aggarwal, Deepti, et al. "IOT based solution for waste management: A review." 2020 8th International Conference on Reliability, Infocom Technologies and Optimization (Trends and Future Directions)(ICRITO). IEEE, 2020.

3. Acharya, Biswa Ranjan, and Pradosh Kumar Gantayat. "Recognition of human unusual activity in surveillance videos." *International Journal of Research and Scientific Innovation (IJRSI)* 2.5 (2015): 18–23.

4. Das, Nivedita, Manjusha Pandey, and Siddharth Swarup Rautaray. "A big step for prediction of HIV/AIDS with big data tools." *Advances in Computer Communication and Computational Sciences: Proceedings of IC4S 2017*, 2.760 (2018): 37.

5. Das, Nivedita, Leena Das, Siddharth Swarup Rautaray, and Manjusha Pandey. "Detection and prevention of HIV AIDS using big data tool." In 2018 3rd International Conference for Convergence in Technology (I2CT), pp. 1–5. IEEE, 2018.

6. Roy, Chandrima, Siddharth Swarup Rautaray, and Manjusha Pandey. "Big data optimization techniques: A survey." *International Journal of Information Engineering & Electronic Business* 10.4 (2018). 41–48.

7. Tripathy, Hrudaya Kumar, et al. "Machine learning on big data: A developmental approach on societal applications." In *Big Data Processing Using Spark in Cloud*. Springer, Singapore, 2019. 143–165.

8. Roy, Chandrima, Manjusha Pandey, and Siddharth Swarup Rautaray. "A proposal for optimization of horizontal scaling in Big data environment." In *Advances in Data and Information Sciences*. Springer, Singapore, 2018. 223–230.

9. Das, Nivedita, Leena Das, Siddharth Swarup Rautaray, and Manjusha Pandey. "Big data analytics for medical applications." *International Journal of Modern Education and Computer Science* 11.2 (2018): 35–42.

10. Bruno, Rodrigo, and Paulo Ferreira. "A study on garbage collection algorithms for big data environments." *ACM Computing Surveys (CSUR)* 51.1 (2018): 1–35.

11. Pandey, Neha, et al. "Smart meter analysis using big data techniques." In *Emerging Research in Computing, Information, Communication and Applications*. Springer, Singapore, 2019. 317–328.

12. Kabi, Kunal, Jyoti Prakash Panigrahi, Bhabani Shankar Prasad Mishra, Manas Kumar Rath, and Satya Ranjan Dash. "Second-generation rearview mirror." In *Smart Intelligent Computing and Applications*. Springer, Singapore, 2020. 603–609.

13. Das, Nivedita, Sandeep Agarwal, Siddharth Swarup Rautaray, and Manjusha Pandey. "Big data approach for epidemiology and prevention of HIV/AIDS." In *Emerging Technologies in Data Mining and Information Security*. Springer, Singapore, 2019. 239–248.

14. Panigrahi, Jyotiprakash, Priyanka Pattnaik, Bibhuti Bhushan Dash, and Satya Ranjan Dash. "Rice quality prediction using computer vision." In 2020 International Conference on Computer Science, Engineering and Applications (ICCSEA). IEEE, 2020. 1–5.

15. Hassanien, Aboul Ella, Nilanjan Dey, and Surekha Borra, eds. *Medical Big Data and Internet of Medical Things: Advances, Challenges and Applications*. CRC Press, Boca Raton, FL, 2018.

16. Anghinolfi, D., M. Paolucci, M. Robba, and A.C. Taramasso. "A dynamic optimization model for solid waste recycling." *Waste Management* 33.2 (2013): 287–296.

17. Hannan, M.A., M. Arebey, R.A. Begum, and H. Basri. "Radio frequency identification (RFID) and communication technologies for solid waste bin truck monitoring system." *Waste Management* 31.12 (2011): 2406–2413.

18. K. Mahajan and J. Chitode, "Zig-Bee based waste bin monitoring system." *International Journal of Engineering Sciences & Research Technology*, 3.2 (Feb 2014), 622–625.

19. V. Bhor, P. Morajkar, and A. Deshpande, "Smart garbage management system." *International Journal of Engineering Research & Technology*, 4.3 (March 2015).

20. Rodic-Wiersma, L. (2013). *Guidelines for National Waste Management Strategies: Moving from Challenges to Opportunities.* UNEP.

21. Srinilta, C., and Kanharattanachai, S. (2019). "Municipal solid waste segregation with CNN." In 2019 5th International Conference on Engineering, Applied Sciences and Technology (ICEAST), Luang Prabang, Laos, 2019. 1–4.

22. Chandramohan, A., J. Mendonca, N. R. Shankar, N. U. Baheti, N. K. Krishnan, and M. S. Suma. "Automated waste segregator." In 2014 Texas Instruments India Educators' Conference (TIIEC), Bangalore, 2014. 1–6.

23. Li, X., Y. Ma, M. Zhang, M. Zhan, P. Wang, X. Lin, J. Yan. "Study on the relationship between waste classification, combustion condition and dioxin emission from waste incineration." *Waste Disposal & Sustainable Energy*, 1.2 (2019): 91–98.

24. Ramos, T., M. I. Gomes, and A. P. Barbosa-Póvoa. "Assessing and improving management practices when planning packaging waste collection systems. Re-sources." *Conservation and Recycling*, 85 (2014): 116–129.

25. Khedikar, A, M. Khobragade, and N. Sawarkar. "Waste management of smart city using IOT." *International Journal of Research in Science and Engineering* 3.2 (2017): 35–38.

26. Schuldt, Christian, Ivan Laptev, and Barbara Caputo. "Recognizing human actions: a local SVM approach." In Proceedings of the 17th International Conference on Pattern Recognition, ICPR 2004, Vol. 3. IEEE, 2004.

27. Roy C., K. Barua, S. Agarwal, M. Pandey, S.S. Rautaray. "Horizontal scaling enhancement for optimized big data processing." In Abraham A., Dutta P., Mandal J., Bhattacharya A., Dutta S. (eds) *Emerging Technologies in Data Mining and Information Security. Advances in Intelligent Systems and Computing*, vol 755. Springer, Singapore. 2019. https://doi.org/10.1007/978-981-13-1951-8_58

28. Hong, Insung, et al. "IoT-based smart garbage system for efficient food waste management." *The Scientific World Journal* 2014 (2014).

29. Agarwal K., E. Maheshwari, C. Roy, M. Pandey, and S.S. Rautray. "Analyzing student performance in engineering placement using data mining." In Chaki N., Devarakonda N., Sarkar A., Debnath N. (eds) *Proceedings of International Conference on Computational Intelligence and Data Engineering. Lecture Notes on Data Engineering and Communications Technologies*, vol. 28. Springer, Singapore. https://doi.org/10.1007/978-981-13-6459-4_18

30. Shyam, Gopal Kirshna, Sunilkumar S. Manvi, and Priyanka Bharti. "Smart waste management using Internet-of-Things (IoT)." In 2017 2nd international conference on computing and communications technologies (ICCCT). IEEE, 2017.

31. Roy, C., M. Pandey, and S. SwarupRautaray. "A proposal for optimization of data node by horizontal scaling of name node using big data tools." In 2018 3rd International Conference for Convergence in Technology (I2CT), 2018. 1–6. doi: 10.1109/I2CT.2018.8529795.

32. Anagnostopoulos, Theodoros, et al. "Challenges and opportunities of waste management in IoT-enabled smart cities: a survey." *IEEE Transactions on Sustainable Computing* 2.3 (2017): 275–289.

33. Medvedev, Alexey, et al. "Waste management as an IoT-enabled service in smart cities." In *Internet of Things, Smart Spaces, and Next Generation Networks and Systems*. Springer, Cham, 2015. 104–115.
34. Roy, Chandrima, and Siddharth Swarup Rautaray. "Challenges and issues of recommender system for big data applications." In *Trends of Data Science and Applications*. Springer, Singapore, 2021. 327–341.
35. Maheshwari, Ekansh, et al. "Prediction of factors associated with the dropout rates of primary to high school students in India using data mining tools." In *Frontiers in Intelligent Computing: Theory and Applications*. Springer, Singapore, 2020. 242–251.

4

Artificial Intelligence and Reducing Food Waste during Harvest and Post-Harvest Processes

Ibrahim A. Abouelsaad, Islam I. Teiba,
Emad H. El-Bilawy, and Islam El-Sharkawy

CONTENTS

DOI: 10.1201/9781003184096-4

4.1 Introduction

Providing food for the growing population worldwide, with many challenges, is of great interest at present [1]. The world population is expected to reach nine billion by the year 2050, and this must be accompanied by an increase of 70% of food [2]. Therefore, many researchers have been concerned with increasing the production of various crops while improving their quality [3–6]. In the meantime, massive quantities of food are being wasted for several reasons. Handling the crops from harvest until they reach the consumer, a quantity of waste occurs that can be estimated at 50%, which might be more in developing countries [7]. It is evident that reducing the loss of crops is of great importance to compensate for food shortages in the world [8]. This can be achieved by improving the crop harvesting, storage, and handling system. Moreover, the use of modern technologies can reduce waste, ensure quality preservation, and help in making the right decisions [9–11].

Understanding the causes of crop waste during harvest and post-harvest is of great importance in identifying problems and reducing waste [2]. Several factors control the preservation of the quantity and quality of crops until they reach the consumer [12, 13]. These factors start from the cultivated field before harvesting. For instance, good follow-up of crops (crop monitoring) before harvest, ensuring that they are free from pests and fungal diseases, and choosing the right time for harvesting are crucial factors in determining crops' quality and increasing shelf life. There may be a tremendous waste of crops during the storage stage due to the lack of appropriate facilities and equipment or the absence of reasonable control over the storage atmosphere (temperature, humidity, ethylene, CO_2, and O_2). Other factors influence the quantity and quality of crops and the level of waste, such as handling, transport packages, transportation, and consumer behavior [2, 7]. The availability of technology that helps producers to monitor, collect information, make decisions, and avoid human errors will reduce food waste [9, 14, 15].

In the agricultural sector, the applications of artificial intelligence (or intelligent agriculture) are in continuous progress as part of the modern technological revolution [9, 16, 17]. The applications of artificial intelligence in agriculture have become of increasing interest due to its potential ability to solve the problems of agricultural labor shortage, climate change, and overpopulation. Agricultural robots are a well-known example of the use of artificial intelligence in the agriculture sector. Many companies are currently developing the programming of these devices and fixing their defects to perform many tasks such as harvesting better and faster than human labor [17]. Crop monitoring is based on deep-learning algorithms and computer vision that analyze the data captured by drones [9, 18]. The machine learning models (Image-Based Predictive Analytics) test massive data amounts that are generated on historical weather patterns, images from drones, the spread of disease, and soil reports to support and improve crop yield. Images taken of different crops under white light or UV-A conditions can determine the ripeness of the fruits and their readiness for harvesting. This helps farmers to separate crops into categories according to maturity before sending them to consumer markets. Overall, artificial intelligence applications can help enhance human lives and solve future problems and challenges [19, 20].

4.2 Food Loss and Food Waste

In the literature, food waste during post-harvest is most likely indicated as food loss [21]. However, as described elsewhere [2], food loss is a term that expresses the decrease in the quantity or quality of food in a way that renders it unfit for consumption. Food waste may occur due to a lack of information, infrastructure, equipment, experiences, and markets. Food waste is a term that includes edible and good food but is discarded due to human behavior [2]. In this chapter, we indicate both food loss and food waste as food waste [21].

4.3 Reasons for Harvest and post-Harvest Loss/Waste

4.3.1 Harvesting

Harvesting is separating the edible portion from the rest of the plant. Manual and mechanical harvesting are standard methods for crop harvest. Harvesting is an important process that determines the size and quality of a crop [22, 23]. The targets of a good harvest are to maximize crop yield and reduce waste while preserving the quality and nutritional value. Logically, inadequate harvesting practices can lead to waste, damage, and reduction of the quality of crops [2, 21]. Crops could be wasted during the harvesting process for several reasons, including leaving the edible parts in the field and exposing them to rotting, plowing the edible parts in the soil, pest attack, improper harvest timing, and mechanical damage during the harvesting process itself [2, 21].

The appropriate time for harvesting is an essential factor determining the amount of waste produced from the economic crops [7]. Nevertheless, harvesting may take place at inappropriate times due to economic reasons or human error in determining the proper times. Unripe vegetables and fruits are susceptible to mechanical damage during harvest, unsuitable for human consumption, and thus are disposed of. Additionally, overripe fruits are subject to mechanical damage during harvest and susceptible to attack by disease and insects.

For cereals, the harvest is also determined by the degree of maturity [2]. Field crops are subjected to drying before harvest to ensure good storage of these crops due to the reduced moisture content. However, at the same time, leaving the field crops for a period without harvest makes them vulnerable to attack by insects, birds, and rodents. On the other hand, an early harvest of crops exposes them to micro-organisms and the decomposition of seeds during storage.

Several studies have determined the appropriate timing for harvesting in different types of vegetables and fruits based on several chemical properties such as the contents of juice, sugar, starch, acidity, oil, protein, and total soluble solids (TSS) [24]. Furthermore, physical properties determine the harvest time for these crops, including size, shape, external and internal color, texture properties, and firmness or tenderness. Both the chemical and physical properties may determine the appropriate harvest time. For example, the appropriate ripening date in tomatoes is determined by fruit size, color, TSS, and PH [25].

4.3.2 Post-Harvest Drying

This process is conducted for field crops to reduce the moisture content before storage. The length of the drying period depends greatly on the weather conditions and the moisture content of the crops after harvest. At this stage, the crops may be vulnerable to attack by birds and rodents. Moreover, large quantities of these crops may be scattered or wasted in the drying places. Accurately adjusting the moisture content of these crops after the drying process is critical to avoid wastage [2]. For instance, if the grains and ears are not sufficiently dried, the high moisture content favors the growth of fungi and molds in storage. On the other hand, if the moisture content of the crops is significantly reduced after the drying process, the grains will become sensitive to cracking and fragmentation, as occurs in rice, which reduces its quality [2].

4.3.3 Packaging

After harvest, crops are often packed to prepare for transportation. Crop packing may be executed improperly, or crops may be deposited in vehicles without any packing. Selecting the appropriate packaging material has a significant role in preserving crops and reducing their waste because these packages help extend the shelf life of vegetables and fruits [7, 23]. On the other hand, low-quality containers that may be purchased for a low price may cause mechanical wounds to the fruits and increase the spread of molds in the stores. The export process requires packages with special characteristics during long-distance transportation, such as gas exchange, preserving moisture content, and reducing the chlorophyll breakdown, as occurs in vegetables.

4.3.4 Transportation

Crops may need to be moved from their production places to the storage places or the markets. During the transport of crops from the fields, they may be subject to wounds (mechanical damage) and consequently fungal infections. Therefore, the quality of these crops may be severely affected, especially in crops that are consumed fresh, such as vegetables and fruits [7, 21]. Accordingly, the global trading system requires fast and efficient transportation that maintains the quantity and quality of crops over long distances.

The method of transportation depends on several factors, such as the nature of those crops, distance, marketing value, and weather conditions (e.g., wind, heat, cold, sunlight, and rainfall) [7, 21]. Moreover, when transporting crops, they may be in a closed space, so other internal factors must be considered, such as air humidity, temperature, mixed loads (types of crops transported together), the rate of moisture loss, and atmospheric gas concentration. These factors directly affect the biological processes within crops [26]. Thus, no doubt, neglecting (or lack of information) these factors greatly influences the shelf life of the vegetable and fruit crops and increases the waste rate before they even reach the market or the desired destination.

4.3.5 Storage

Crops may be stored for several reasons. Crop storage is necessary to increase the supply of crops in times of scarcity and to avoid a shortage of supplies in the market. Also, crops may be stored due to the availability of water and the appropriate weather conditions for the occurrence of abundant agricultural production in a certain period of the year without the rest of the periods. Economic conditions also control the farmers, where the farmer may resort to storing crops due to the low prices and hoping to resell them at a better price after a suitable storage period.

Storage is a method of preserving the quantity and quality of crops with less spoilage and waste for a period longer than the average shelf life for those crops [2]. Crop storage depends mainly on scientific methods and principles; thus, paying attention to these principles is the basis for the success of the storage process [27]. In closed storage conditions, control of hygiene, humidity, temperature, atmospheric gas concentration (e.g., CO_2, O_2, and ethylene), and ventilation dramatically contributes to the success of long-term storage [28]. Also, controlling pests and avoiding mold growth reduce food waste effectively. Therefore, the lack of adequate storage facilities in developing countries may be the leading cause of crop waste after harvest [2].

4.3.6 Marketing

The marketing process depends heavily on the previous stages [12]. Of course, high-quality crops have a good chance of selling compared to low-quality crops that carry insects and defects. Also, choosing the appropriate time to offer crops for sale and choosing the period of consumer demand during the year has a great role in avoiding the waste of these crops. Crops also differ in their ability to stay in the market, as vegetable and fruit crops must be marketed quickly, especially in the summer. Farmers must be aware of all these factors and have information for the marketing process's success to avoid crops being wasted by not reaching consumers.

4.3.7 Consumer

The consumption stage is the arrival of the crops to the consumer after they are purchased from the market. In a study carried out by FAO [29], vegetables and fruits account for 39 % of total household waste. There are many reasons for the occurrence of such a large amount of waste, such as over-purchasing without planning, home storage facilities, income, and socio-cultural factors (e.g., gender and lifestyle) [13]. In general, there are noticeable differences between countries and their economic power in the causes of agricultural crop waste. In developing countries, the waste is mainly due to the lack of capacity for storage and cooling in homes. However, waste is primarily due to consumer behavior, safety policies, and quality standard requirements in developed countries [7].

4.4 Artificial Intelligence Applications for Reducing Crop Loss/Waste

4.4.1 Crop Monitoring during Production

Sensors and imaging devices have helped producers in the agricultural sector to reduce yield losses and increase production. High-resolution cameras and sensors mounted on unmanned aerial vehicles (UAV) are the eyes of the farmer on the farm to obtain, analyze, and test data [9, 14]. In fact, the use of aerial survey methods is not new in the field of agriculture. A decade ago, satellites were used to study forests and huge agricultural areas. However, the use of drones allowed a new level of accuracy and made it easier to collect and analyze data. For instance, the UAV provides high-resolution and high-quality images independent of satellite location because the images are taken from a level close to farmland [18, 30].

In studies on the feasibility and challenges of using drones in agriculture, the ability of these devices to help farmers in increasing yield and reduce wastage has been demonstrated [31]. This was performed by early detection of problems such as pests and dehydration using special cameras. Monitoring crops (health status) using a multispectral imaging system installed on a drone was achieved [32]. A microcontroller with two cameras was used in that system; the first is a standard RGB camera, and the other is sensitive to infrared radiation. The system exhibits images used by software to compute the normalized difference vegetation index (NDVI). Furthermore, sensors mounted on UAV with the optimal procedures for surveying, data collection, and analysis are continuously developed and tested for precision farming applications [33].

4.4.1.1 Weather Prediction

The crops interact with the surrounding weather conditions [34]. Predicting weather conditions such as rainfall, humidity, average sunshine, and drought guide farmers to optimally plan agricultural and harvesting practices [35]. Weather conditions also directly control the spread of disease and insects that are moved with crops to the storages and markets and reduce their quality, thus increasing waste. Table 4.1 lists various artificial intelligence technologies used in weather prediction.

4.4.1.2 Disease Detection

Using artificial intelligence techniques (e.g., pattern recognition and machine learning), it is now possible to detect early pathological and insect infections before their spread on the farm (see Table 4.2) [42]. Early elimination of these diseases and insects and preventing their spread increases the quality of crops and thus increases marketing opportunities and consumer demand for them. Also, reducing disease and insect injuries in the fields significantly reduces their transport to storage areas after harvest.

4.4.1.3 Nutrient Management

To obtain good quality traits of crops, adequate quantities of nutrients are required in the soil [52–54]. A crop quality management application such as

TABLE 4.1

Summary of Artificial Intelligence Technologies Used in Weather Prediction

Technique	Application	References
Genetic algorithm	Rainfall prediction	[36]
Ensemble learning	Rainfall prediction	[37]
Regression	Rainfall prediction	[38]
Regression	Drought forecasting	[39]
Artificial neural network	Climate change impact	[40]
Decision tree	Climate change impact	[40]
Instance-based learning	Evapotranspiration estimation	[41]

TABLE 4.2

Summary of Artificial Intelligence Technologies Used in Disease Detection

Technique	Application	References
Artificial neural network	Disease detection	[43]
Ensemble learning	Wheat yellow rust monitoring	[44]
Instance-based learning	Leaf disease detection	[45]
Support-vector machine	Leaf disease detection	[45]
Artificial neural network	Grape leaf disease detection	[46]
Image processing, K-mean clustering	Disease detection in Malus domestica	[47]
Machine vision and image processing techniques	Identification of citrus disease	[48]
Computer vision and image processing algorithms	Detection of fungal diseases in fruits and vegetables	[49]
Artificial neural network	Detect rice leaf disease	[50]
Fuzzy logic with k-means segmentation	Disease severity of rice crop	[51]

TABLE 4.3

Summary of Nutrient Management Achieved by Artificial Intelligence

Technique	Application	References
Artificial neural network	Nitrogen status estimation	[56]
Artificial neural network	Prediction of soil fertility for several nutrients	[57]
Deep learning	Prediction of soil fertility for several nutrients	[57]
Hyperspectral frame camera	Monitoring the leaf nitrogen level in rice	[58]
Digital photography from model aircraft	Remote sensing of crop nitrogen status	[59]
Management-oriented modeling	Minimizes nitrate leaching	[60]
Artificial neural network	Estimate soil nutrients after erosion	[61]
Remote sensing, decision tree	Nitrogen application rates in corn	[62]

site-specific nutrient management is an optimal method for the detection and classification of crop quality parameters [9]. This, of course, helps to increase the market value of crops and reduce loss or waste [55]. Table 4.3 shows the different artificial intelligence technologies used for nutrient management.

4.4.1.4 Crop Yield Prediction

Predicting crop yield helps effectively in managing crop production and determining many plans for marketing. For example, farm inputs, such as fertilizers and equipment required for harvesting, can be determined according to soil and crop needs, thus increasing field efficiency [63]. The forecasting of crop yield also helps in determining the areas required for storage, the number of packages, and the expected economic return for those crops (see Table 4.4).

4.4.2 Demand Prediction and Production Planning

The main objective of this stage is to predict the demand for food (crops) and plan ahead for agricultural production. Predicting the size of food needs greatly helps to avoid overproduction and reduce crop losses and also helps in the optimal use of resources [72]. The use of big data analytics and ML algorithms helps reduce setup time and improve the sense of demand [9] (see Table 4.5).

4.4.3 Distribution and Transportation

At this stage, studies are concerned with analyzing road problems and reducing the damage of agricultural products while preserving their quality [9, 14]. These studies also contribute to the evaluation and analysis of distribution problems with many applications for distribution scheduling, identifying the storage location, storage management, dynamic allocation,

TABLE 4.4

Crop Yield Prediction Achieved by Artificial Intelligence

Technique	Application	References
Artificial neural network	Prediction of cotton yield	[64]
Artificial neural network	Prediction of corn yield	[65]
Regression	Forecasting yield of maturing rice	[66]
Fuzzy cognitive map learning	Prediction of apples yield	[67]
Regression	Pre-harvest forecast of sugarcane yield	[68]
Decision tree algorithms	Soybean productivity	[69]
Artificial neural network approach	Agricultural crop yield estimating	[70]
Artificial neural network	Rice yield prediction	[71]

TABLE 4.5

Artificial Intelligence for Demand Prediction and Production Planning

Technique	Application	References
Artificial neural network	Modeling consumer's behavior for packed vegetable	[73]
Artificial neural network	Demand forecasting for the foodstuff retail segment	[74]
Genetic algorithm	Food supply chain management	[75]
Genetic algorithm	Optimization of economics and environmental life cycle assessment in oilseed production	[76]
Genetic algorithm	Environmental life cycle assessment in paddy production	[77]
Genetic algorithm	Estimation of food shelf life	[78]

developing local food supply chains, estimation of product shelf life, and predicting supply chain risks under uncertainties (see Table 4.6) [9, 18].

4.4.4 Consumer Analytics

Providing accurate data on consumer needs and the connected trends will assist producers in setting priorities, better planning ahead, and saving resources. Machine learning techniques (e.g., Bayesian network, clustering, artificial neural network, and support vector machine) are used to predict consumer needs and behavior in the buying process (see Table 4.7).

TABLE 4.6

Summary of Artificial Intelligence Technologies Used in Distribution and Transportation

Technique	Application	References
Regression	Quality assurance in sustainable food supply chain.	[79]
Clustering	Distribution planning for fruit-and-vegetable cold chains.	[80]
Genetic algorithm	Evaluate vehicle routing problem in the delivery of fresh agricultural products.	[81]
Genetic algorithm	Optimizing integrated inventory policy for a multi-stage supply. Chain	[82]
Genetic algorithm	Cost-optimization for fresh food quality and transportation	[83]
Genetic algorithm	Minimize food damage and estimate vehicle routing problem and considering road conditions	[84]
Regression	A logistic network to optimize the development of local food system with safety and sustainability	[85]
Regression	Monitor the crops stored in the supply chain inventories and provide status about their quality	[86]

TABLE 4.7

Different Artificial Intelligence Applications Used for Consumer Needs and Behavior

Technique	Application	References
Artificial neural network	Classifying consumer purchasing decision	[87]
Bayesian network	Analyzing organic food buyers	[88]
Clustering	Analysis of the reasons for buying organic rice	[89]
Support vector machines	Forecasting consumer healthy choices regarding wheat flour	[90]
Artificial neural network	Predicting consumer demand and buying behavior	[91]
Support vector machines	Customer feedback	[92]
Bayesian network	Predicting the consumers buying behavior	[73]
Wavelet neural networks	Forecasting soybean sack price and the customer demand	[93]

4.4.5 Harvesting Robots

Robotics is used in many different fields such as the automotive industry, medicine, and military equipment. In the agricultural sector, robots supported by artificial intelligence were used to accomplish many heavy and labor-intensive tasks such as planting seeds, spraying, and harvesting [17, 94]. The acute shortage of agricultural labor and the increase in migration from the countryside to the cities have demonstrated the importance of harvest robots in increasing agricultural production, reducing waste, and improving sustainable agriculture [17].

The idea of using this technology began in a cotton gin device in 1794 by the inventor Eli Whitney (1765–1825) [14]. This machine was used to quickly separate the seeds from the cotton fiber, which revolutionized cotton production at that time. This was followed by a great development in agricultural robots in planting seeds and removing seeds employing infrared light [95]. Harvest robots operate in a very complex environment due to the large differences between crops in size, shape, color, and texture of fruits. Therefore, harvest robots that depend on machine vision should have the ability to collect information and adapt to detect the target of different crops and learn independently [96]. The robots, as intelligent automated machines, should have intelligent reasoning for decision making and a network transmission function for sending the fruit images to a data center or server [17, 96, 97]. Table 4.8 lists various agricultural harvesting robots.

4.5 Conclusions

Agricultural production goes through many stages, starting from growing crops in the fields, through to harvesting and trading them until reaching the final consumer. During these many stages, crops are susceptible to spoilage, waste, and reduced quality. With the availability of artificial intelligence technology and the introduction of its application in the agricultural sector, it is possible to increase agricultural production, improve quality, and reduce waste. This is done by monitoring crops as they grow, tracking disease infections, crop needs of elements, and forecasting weather conditions. The use of robots in harvesting may replace many arduous tasks for farmers and allow crops to be classified according to their quality and degree of maturity. The applications of artificial intelligence also provide the opportunity to monitor crops in stores, and determine market requirements and consumer trends, which will greatly help in planning in advance for production and identifying priorities. The technology of artificial intelligence and its application in the agricultural sector is a continuous progression to face many challenges such as high costs, the accuracy of results, low speed of harvesting, and the

TABLE 4.8

Robots Used in Harvesting

Recognition Method	Fruits Recognition Accuracy (%)	Time Required to Harvest One Fruit (s)	References
Vision-Based Target (Multi box detector plus a stereo camera)	90	16	[98]
Stereoscopic 3D vision sensing to recognize oil palm fresh fruits	65–70	-	[99]
Stereoscopic vision (stereo camera) for apples location	89.5	-	[100]
Laser range finder (test the role of light to get information) for tomato cluster harvesting	-	-	[101]
A multispectral imaging analysis for citrus fruit detection	88–93	-	[102]
Feature images fusion for robust tomato detection	93	-	[103]
Deep detection network (in-depth learning), to locate the iceberg lettuce	91	31.7	[104]
Binocular stereo vision for clustered tomatoes detection	87.9	-	[105]
Vision-based determination of spatial information (binocular stereo vision) for grape clusters harvesting	87	-	[106]
Double otsu algorithm (multiple color targets) for litchi fruit detection	95	-	[107]
A vision servo system (three cameras to make the visual servo system) for sweet pepper harvest	82.16	51.1	[108]
Visual system by stereoscopic vision plus deep neural network for kiwi fruit picking	76.1	5.5	[109]
Citrus recognition by shape analysis method and fruit detection algorithm	90	-	[110]
Fruit detection from images by deep learning	96.3	-	[111]
Fruit geometry and color space for ripe tomato detection	96	-	[112]

ability to apply to different types of crops and under different environmental conditions.

References

1. Dossou, S.; Dawood, M.A.O.; Zaineldin, A.I.; Abouelsaad, I.A.; Mzengereza, K.; Shadrack, R.S.; Zhang, Y.; El-Sharnouby, M.; Ahmed, H.A.; El Basuini, M.F.

Dynamical hybrid system for optimizing and controlling efficacy of plant-based protein in aquafeeds. *Complexity* 2021, *2021*, 1–7, doi:10.1155/2021/9957723.

2. Kiaya, V. Post-harvest losses and strategies to reduce them. In *Technical Paper on Postharvest Losses, Action Contre la Faim (ACF)*; 2014, *25*, 1–25.

3. Eickhout, B.; Bouwman, A.F.; van Zeijts, H. The role of nitrogen in world food production and environmental sustainability. *Agric. Ecosyst. Environ.* 2006, *116*, 4–14, doi:10.1016/j.agee.2006.03.009.

4. Sary Hassan, M.B.; Ibrahim Ali, A.A. The combined use of beneficial soil micro-organisms enhanced the growth and efficiently reduced lead content in leaves of lettuce (Lactuca sativa L.) Plant under Lead Stress. *Alexandria J. Agric. Sci.* 2019, *64*, 41–51, doi:10.21608/alexja.2019.41854.

5. Brengi, S.H.; Khedr, A.A.E.M.; Abouelsaad, I.A. Effect of melatonin or cobalt on growth, yield and physiological responses of cucumber (Cucumis sativus L.) plants under salt stress. *J. Saudi Soc. Agric. Sci.* 2021, *21*(1), 51–60. doi:10.1016/J.JSSAS.2021.06.012.

6. Abouelsaad, I.; Renault, S. Enhanced oxidative stress in the jasmonic acid-deficient tomato mutant def-1 exposed to NaCl stress. *J. Plant Physiol.* 2018, *226*, 136–144, doi:10.1016/J.JPLPH.2018.04.009.

7. Yanık, D.K.; Elik, A.; Yanik, D.K.; Guzelsoy, N.A.; Yavuz, A.; Gogus, F. Strategies to reduce post-harvest losses for fruits and vegetables. *Strategies* 2019, *5*(3), 29–39. doi:10.7176/JSTR/5-3-04.

8. Gustafsson, U.; Wills, W.; Draper, A. Food and public health: Contemporary issues and future directions. *Crit. Public Health* 2011, *21*, 385–393.

9. Sharma, R.; Kamble, S.S.; Gunasekaran, A.; Kumar, V.; Kumar, A. A systematic literature review on machine learning applications for sustainable agriculture supply chain performance. *Comput. Oper. Res.* 2020, *119*, 104926, doi:10.1016/j.cor.2020.104926.

10. Zidan, M.; Abdel-Aty, A.H.; Nguyen, D.M.; Mohamed, A.S.A.; Al-Sbou, Y.; Eleuch, H.; Abdel-Aty, M. A quantum algorithm based on entanglement measure for classifying Boolean multivariate function into novel hidden classes. *Results Phys.* 2019, *15*, 102549, doi:10.1016/j.rinp.2019.102549.

11. Zidan, M.; Abdel-Aty, A.H.; El-Sadek, A.; Zanaty, E.A.; Abdel-Aty, M. Low-cost autonomous perceptron neural network inspired by quantum computation. In Proceedings of the AIP Conference Proceedings; American Institute of Physics Inc.. 2017; Vol. 1905, p. 020005.

12. Shamsher, M.; Mohammed, A.; Siddiqui, W. *Postharvest Quality Assurance of Fruits Practical Approaches for Developing Countries*; Springer, Cham. 2015. https://doi.org/10.1007/978-3-319-21197-8

13. Porat, R.; Lichter, A.; Terry, L.A.; Harker, R.; Buzby, J. Postharvest losses of fruit and vegetables during retail and in consumers' homes: Quantifications, causes, and means of prevention. *Postharvest Biol. Technol.* 2018, *139*, 135–149.

14. Talaviya, T.; Shah, D.; Patel, N.; Yagnik, H.; Shah, M. Implementation of artificial intelligence in agriculture for optimisation of irrigation and application of pesticides and herbicides. *Artif. Intell. Agric.* 2020, *4*, 58–73, doi:10.1016/j.aiia.2020.04.002.

15. Liu, W.; Wu, Q.; Shen, J.; Zhao, J.; Zidan, M.; Tong, L. An optimized quantum minimum searching algorithm with sure-success probability and its experiment simulation with Cirq. *J. Ambient Intell. Humaniz. Comput.* 2021, *1*, 3, doi:10.1007/s12652-020-02840-z.

16. Zidan, M. A novel quantum computing model based on entanglement degree. *Mod. Phys. Lett. B* 2020, *34*, doi:10.1142/S0217984920504011.
17. Tang, Y.; Chen, M.; Wang, C.; Luo, L.; Li, J.; Lian, G.; Zou, X. Recognition and Localization Methods for Vision-Based Fruit Picking Robots: A Review. *Front. Plant Sci.* 2020, *11*.
18. Pathan, M.; Patel, N.; Yagnik, H.; Shah, M. Artificial cognition for applications in smart agriculture: A comprehensive review. *Artif. Intell. Agric.* 2020, *4*, 81–95, doi:10.1016/j.aiia.2020.06.001.
19. Sagheer, A.; Zidan, M.; Abdelsamea, M.M. A novel autonomous perceptron model for pattern classification applications. *Entropy* 2019, *21*, 763, doi:10.3390/e21080763.
20. Zidan, M.; Eleuch, H.; Abdel-Aty, M. Non-classical computing problems: Toward novel type of quantum computing problems. *Results Phys.* 2021, *21*, 103536, doi:10.1016/j.rinp.2020.103536.
21. Parfitt, J.; Barthel, M.; MacNaughton, S. Food waste within food supply chains: Quantification and potential for change to 2050. *Philos. Trans. R. Soc. B Biol. Sci.* 2010, *365*, 3065–3081.
22. Negi, S. Supply chain efficiency: An insight from fruits and vegetables sector in India. *J. Oper. Supply Chain Manag.* 2014, *7*, 154, doi:10.12660/joscmv7n2p154-167.
23. Ahmad, M.S.; Siddiqui, M.W.; Ahmad, M.S.; Siddiqui, M.W. Introduction. In *Postharvest Quality Assurance of Fruits*; Springer International Publishing; 2015; pp. 1–5.
24. Fawole, O.A.; Opara, U.L. Developmental changes in maturity indices of pomegranate fruit: A descriptive review. *Sci. Hortic.* 2013, *159*, 152–161.
25. Okiror, P.; Lejju, J.B.; Bahati, J.; Rugunda, G.K.; Sebuuwufu, C.I. Maturity indices for tomato (Solanum lycopersicum L.), cv. Ghalia 281 in Central Uganda. *African J. Agric. Res.* 2017, *12*, 1196–1203, doi:10.5897/AJAR2017.12150.
26. Hertog, M.L.A.T.M.; Uysal, I.; McCarthy, U.; Verlinden, B.M.; Nicolaï, B.M. Shelf life modelling for first-expired-first-out warehouse management. *Philos. Trans. R. Soc. A Math. Phys. Eng. Sci.* 2014, *372*.
27. Ullah Khan, I.; Hafiz Dzarfan Othman, M.; Hashim, H.; Matsuura, T.; Ismail, A.F.; Rezaei-DashtArzhandi, M.; Wan Azelee, I. Biogas as a renewable energy fuel: A review of biogas upgrading, utilisation and storage. *Energy Convers. Manag.* 2017, *150*, 277–294.
28. Pessu, P.O.; Agoda, S.; Isong, I.U.; Ikotun, I. The concepts and problems of post-harvest food losses in perishable crops. *African J. Food Sci.* 2011, *5*, 603–613.
29. Global food losses and food waste Available online: http://www.fao.org/3/mb060e/mb060e00.htm (accessed on May 9, 2021).
30. Gongal, A.; Amatya, S.; Karkee, M.; Zhang, Q.; Lewis, K. Sensors and systems for fruit detection and localization: A review. *Comput. Electron. Agric.* 2015, *116*, 8–19.
31. Reinecke, M.; Prinsloo, T. The influence of drone monitoring on crop health and harvest size. In Proceedings of the 2017 1st International Conference on Next Generation Computing Applications, NextComp 2017; Institute of Electrical and Electronics Engineers Inc., 2017; pp. 5–10.
32. De Oca, A.M.; Arreola, L.; Flores, A.; Sanchez, J.; Flores, G. Low-cost multispectral imaging system for crop monitoring. In Proceedings of the 2018 International Conference on Unmanned Aircraft Systems, ICUAS 2018; Institute of Electrical and Electronics Engineers Inc., 2018; pp. 443–451.

33. Sona, G.; Passoni, D.; Pinto, L.; Pagliari, D.; Masseroni, D.; Ortuani, B.; Facchi, A. UAV multispectral survey to map soil and crop for precision farming applications. In Proceedings of the International Archives of the Photogrammetry, Remote Sensing and Spatial Information Sciences - ISPRS Archives; International Society for Photogrammetry and Remote Sensing; 2016; Vol. 2016, January, pp. 1023–1029.

34. Abouelsaad, I. Salinity tolerance of tomato plants: the role of jasmonic acid and root ammonium transporters; 2017. PhD Thesis. University of Manitoba. http://hdl.handle.net/1993/32070

35. Bendre, M.R.; Thool, R.C.; Thool, V.R. Big data in precision agriculture through ICT: Rainfall prediction using neural network approach. In Proceedings of the Advances in Intelligent Systems and Computing; Springer Verlag; 2016; Vol. 438, pp. 165–175.

36. Wahyuni, I.; Firdaus Mahmudy, W. Rainfall Prediction in Tengger Indonesia Using Hybrid Tsukamoto FIS and Genetic Algorithm Fuzzy Time Series View project. *Artic. J. ICT Res. Appl.* 2017, doi:10.5614/itbj.ict.res.appl.2017.11.1.3.

37. Ali, M.; Deo, R.C.; Downs, N.J.; Maraseni, T. Multi-stage committee based extreme learning machine model incorporating the influence of climate parameters and seasonality on drought forecasting. *Comput. Electron. Agric.* 2018, *152*, 149–165, doi:10.1016/j.compag.2018.07.013.

38. Cramer, S. *New Genetic Programming Methods for Rainfall Prediction and Rainfall Derivatives Pricing.* 2017. PhD Thesis. University of Kent (KAR id:69471). https://kar.kent.ac.uk/69471/

39. Belayneh Abcdef, A.; Adamowski Abcdef, J. Environmental Engineering in Agriculture. *J. Water L. Dev.* 2013, 3–12, doi:10.2478/jwld-2013-0001.

40. Crane-Droesch, A. Machine learning methods for crop yield prediction and climate change impact assessment in agriculture. *Environ. Res. Lett.* 2018, *13*, 114003, doi:10.1088/1748-9326/aae159.

41. Saggi, M.K.; Jain, S. Reference evapotranspiration estimation and modeling of the Punjab Northern India using deep learning. *Comput. Electron. Agric.* 2019, *156*, 387–398, doi:10.1016/j.compag.2018.11.031.

42. Lee, W.S.; Alchanatis, V.; Yang, C.; Hirafuji, M.; Moshou, D.; Li, C. Sensing technologies for precision specialty crop production. *Comput. Electron. Agric.* 2010, *74*, 2–33.

43. Kamilaris, A.; Prenafeta-Boldú, F.X. Deep learning in agriculture: A survey. *Comput. Electron. Agric.* 2018, *147*, 70–90.

44. Su, J.; Liu, C.; Coombes, M.; Hu, X.; Wang, C.; Xu, X.; Li, Q.; Guo, L.; Chen, W.H. Wheat yellow rust monitoring by learning from multispectral UAV aerial imagery. *Comput. Electron. Agric.* 2018, *155*, 157–166, doi:10.1016/j.compag.2018.10.017.

45. Pantazi, X.E.; Moshou, D.; Tamouridou, A.A. Automated leaf disease detection in different crop species through image features analysis and One Class Classifiers. *Comput. Electron. Agric.* 2019, *156*, 96–104, doi:10.1016/j.compag.2018.11.005.

46. Kakade, N.R.; Ahire, D.D. Real time grape leaf disease detection. *Int. J. Adv. Res. Innov. Ideas Educ. (IJARIIE)* 2015, 1(4), 1.

47. Bashir, S.; Sharma, N. *Remote Area Plant Disease Detection Using Image Processing;* Vol. 2.

48. Pydipati, R.; Burks, T.F.; Lee, W.S. Identification of citrus disease using color texture features and discriminant analysis. *Comput. Electron. Agric.* 2006, *52*, 49–59, doi:10.1016/j.compag.2006.01.004.

49. Pujari, J.D.; Yakkundimath, R.; Byadgi, A.S. Image processing based detection of fungal diseases in plants. In Proceedings of the Procedia Computer Science; Elsevier B.V.; 2015; Vol. 46, pp. 1802–1808.

50. Ghyar, B.S.; Birajdar, G.K. Computer vision based approach to detect rice leaf diseases using texture and color descriptors. In Proceedings of the Proceedings of the International Conference on Inventive Computing and Informatics, ICICI 2017; Institute of Electrical and Electronics Engineers Inc.; 2018; pp. 1074–1078.

51. Sethy, P.K.; Negi, B.; Barpanda, N.K.; Behera, S.K.; Rath, A.K. Measurement of disease severity of rice crop using machine learning and computational intelligence. In *SpringerBriefs in Applied Sciences and Technology*; Springer Verlag; 2018; pp. 1–11.

52. Brengi, S.H.; Abouelsaad, I.A. The role of different nitrogen sources combined with foliar applications of molybdenum, selenium or sucrose in improving growth and quality of edible parts of spinach (Spinacia oleracea L.). *Alexandria Sci. Exch. J.* 2019, *40*, 156–168, doi:10.21608/asejaiqjsae.2019.29731.

53. Abouelsaad, I.; Weihrauch, D.; Renault, S. Effects of salt stress on the expression of key genes related to nitrogen assimilation and transport in the roots of the cultivated tomato and its wild salt-tolerant relative. *Sci. Hortic.* 2016, *211*, 70–78, doi:10.1016/j.scienta.2016.08.005.

54. Hassan, S.; AbouelSaad, I.A.; Roshdy, A.H. Growth, yield and nutrient contents of garlic as affected by bio-inoculants and mineral fertilizers. *J. Agric. Env. Sci, Damanhour Univ.* 2018, *17*, 1–19.

55. Abbal, P.; Sablayrolles, J.M.; Matzner-Lober, É.; Boursiquot, J.M.; Baudrit, C.; Carbonneau, A. A decision support system for vine growers based on a Bayesian network. *J. Agric. Biol. Environ. Stat.* 2016, *21*, 131–151, doi:10.1007/s13253-015-0233-2.

56. Chlingaryan, A.; Sukkarieh, S.; Whelan, B. Machine learning approaches for crop yield prediction and nitrogen status estimation in precision agriculture: A review. *Comput. Electron. Agric.* 2018, *151*, 61–69.

57. Sirsat, M.S.; Cernadas, E.; Fernández-Delgado, M.; Barro, S. Automatic prediction of village-wise soil fertility for several nutrients in India using a wide range of regression methods. *Comput. Electron. Agric.* 2018, *154*, 120–133, doi:10.1016/j.compag.2018.08.003.

58. Zheng, H.; Zhou, X.; Cheng, T.; Yao, X.; Tian, Y.; Cao, W.; Zhu, Y. Evaluation of a UAV-based hyperspectral frame camera for monitoring the leaf nitrogen concentration in rice. In Proceedings of the International Geoscience and Remote Sensing Symposium (IGARSS); Institute of Electrical and Electronics Engineers Inc.; 2016; Vol. 2016-November, pp. 7350–7353.

59. Hunt, E.R.; Cavigelli, M.; Daughtry, C.S.T.; McMurtrey, J.E.; Walthall, C.L. Evaluation of digital photography from model aircraft for remote sensing of crop biomass and nitrogen status. *Precis. Agric.* 2005, *6*, 359–378, doi:10.1007/s11119-005-2324-5.

60. Li, M.; Yost, R.S. Management-oriented modeling: Optimizing nitrogen management with artificial intelligence. *Agric. Syst.* 2000, *65*, 1–27, doi:10.1016/S0308-521X(00)00023-8.

61. Kim, M.; Gilley, J.E. Artificial Neural Network estimation of soil erosion and nutrient concentrations in runoff from land application areas. *Comput. Electron. Agric.* 2008, *64*, 268–275, doi:10.1016/j.compag.2008.05.021.

62. Waheed, T.; Bonnell, R.B.; Prasher, S.O.; Paulet, E. Measuring performance in precision agriculture: CART-A decision tree approach. *Agric. Water Manag.* 2006, *84*, 173–185, doi:10.1016/j.agwat.2005.12.003.

63. Elavarasan, D.; Vincent, D.R.; Sharma, V.; Zomaya, A.Y.; Srinivasan, K. Forecasting yield by integrating agrarian factors and machine learning models: A survey. *Comput. Electron. Agric.* 2018, *155*, 257–282.

64. Haghverdi, A.; Washington-Allen, R.A.; Leib, B.G. Prediction of cotton lint yield from phenology of crop indices using artificial neural networks. *Comput. Electron. Agric.* 2018, *152*, 186–197, doi:10.1016/j.compag.2018.07.021.

65. Khanal, S.; Fulton, J.; Klopfenstein, A.; Douridas, N.; Shearer, S. Integration of high resolution remotely sensed data and machine learning techniques for spatial prediction of soil properties and corn yield. *Comput. Electron. Agric.* 2018, *153*, 213–225, doi:10.1016/j.compag.2018.07.016.

66. Shibayama, M.; Akiyama, T. Estimating grain yield of maturing rice canopies using high spectral resolution reflectance measurements. *Remote Sens. Environ.* 1991, *36*, 45–53, doi:10.1016/0034-4257(91)90029-6.

67. Papageorgiou, E.I.; Aggelopoulou, K.D.; Gemtos, T.A.; Nanos, G.D. Yield prediction in apples using Fuzzy Cognitive Map learning approach. *Comput. Electron. Agric.* 2013, *91*, 19–29, doi:10.1016/j.compag.2012.11.008.

68. Priya, S.R.K.; Suresh, K.K. *A Study on Pre-harvest Forecast of Sugarcane Yield Using Climatic Variables*; 2009; Vol. 7.

69. Suraparaju, V.; Veenadhari, S.; Mishra, B.; Singh, C.D. Soybean productivity modelling using decision tree algorithms. *Int. J. Comput. Appl.* 2011, *27*, 975–8887, doi:10.13140/RG.2.1.3852.1846.

70. Dahikar, S.S.; Rode, S.V. *Agricultural Crop Yield Prediction Using Artificial Neural Network Approach*; 2014; Vol. 2.

71. Ji, B.; Sun, Y.; Yang, S.; Wan, J. Artificial neural networks for rice yield prediction in mountainous regions. In Proceedings of the Journal of Agricultural Science; Cambridge University Press; 2007; Vol. 145, pp. 249–261.

72. Feng, Q.; Shanthikumar, J.G. How research in production and operations management may evolve in the era of big data. *Prod. Oper. Manag.* 2018, *27*, 1670–1684, doi:10.1111/poms.12836.

73. Borimnejad, V.; Eshraghi Samani, R. Modeling consumer's behavior for packed vegetable in "Mayadin management organization of Tehran" using artificial neural network. *Cogent Bus. Manag.* 2016, 3.

74. da Veiga, C.P.; da Veiga, C.R.P.; Puchalski, W.; Coelho, L. dos S.; Tortato, U. Demand forecasting based on natural computing approaches applied to the foodstuff retail segment. *J. Retail. Consum. Serv.* 2016, *31*, 174–181, doi:10.1016/j.jretconser.2016.03.008.

75. Sitek, P.; Wikarek, J.; Nielsen, P. A constraint-driven approach to food supply chain management. *Ind. Manag. Data Syst.* 2017, *117*, 2115–2138, doi:10.1108/IMDS-10-2016-0465.

76. Mousavi-Avval, S.H.; Rafiee, S.; Sharifi, M.; Hosseinpour, S.; Notarnicola, B.; Tassielli, G.; Renzulli, P.A. Application of multi-objective genetic algorithms for optimization of energy, economics and environmental life cycle assessment in oilseed production. *J. Clean. Prod.* 2017, *140*, 804–815, doi:10.1016/j.jclepro.2016.03.075.

77. Nabavi-Pelesaraei, A.; Rafiee, S.; Mohtasebi, S.S.; Hosseinzadeh-Bandbafha, H.; Chau, K. wing Energy consumption enhancement and environmental life cycle

assessment in paddy production using optimization techniques. *J. Clean. Prod.* 2017, *162*, 571–586, doi:10.1016/j.jclepro.2017.06.071.

78. Larsen, R.A.; Schaalje, G.B.; Lawson, J.S. Food shelf life: Estimation and optimal design. *J. Stat. Comput. Simul.* 2010, *80*, 143–157, doi:10.1080/00949650802549135.

79. Ting, S.L.; Tse, Y.K.; Ho, G.T.S.; Chung, S.H.; Pang, G. Mining logistics data to assure the quality in a sustainable food supply chain: A case in the red wine industry. *Int. J. Prod. Econ.* 2014, *152*, 200–209, doi:10.1016/j.ijpe.2013.12.010.

80. Hsiao, Y.H.; Chen, M.C.; Lu, K.Y.; Chin, C.L. Last-mile distribution planning for fruit-and-vegetable cold chains. *Int. J. Logist. Manag.* 2018, *29*, 862–886, doi:10.1108/IJLM-01-2017-0002.

81. Qiang, L.; Qiang, L.; Jiuping, X. *A Study on Vehicle Routing Problem in the Delivery of Fresh Agricultural Products under Random Fuzzy Environment. Int. J. Inf. Manag. Sci.* 2008, *19*(4), 673–690.

82. Dolgui, A.; Tiwari, M.K.; Sinjana, Y.; Kumar, S.K.; Son, Y.J. Optimising integrated inventory policy for perishable items in a multi-stage supply chain. *Int. J. Prod. Res.* 2018, *56*, 902–925, doi:10.1080/00207543.2017.1407500.

83. Nakandala, D.; Lau, H.; Zhang, J. Cost-optimization modelling for fresh food quality and transportation. *Ind. Manag. Data Syst.* 2016, *116*, 564–583, doi:10.1108/IMDS-04-2015-0151.

84. Padilla, M.P.B.; Canabal, P.A.N.; Pereira, J.M.L.; Riaño, H.E.H. Vehicle routing problem for the minimization of perishable food damage considering road conditions. *Logist. Res.* 2018, *11*, 1–18, doi:10.23773/2018_2.

85. Saetta, S.A.; Caldarelli, V.; Tiacci, L.; Lerche, N.; Geldermann, J. A logistic network to harmonise the development of local food system with safety and sustainability. *Int. J. Integr. Supply Manag.* 2015, *9*, 307–328, doi:10.1504/IJISM.2015.070530.

86. Kumar Mishra, C. Post-harvest crop management system using IoT and AI. *Int. J. Adv. Res. Dev.* 2019, *4*(5), 42–44.

87. Lilavanichakul, A.; Chaveesuk, R.; Kessuvan, A. Classifying Consumer Purchasing Decision for Imported Ready-to-Eat Foods in China Using Comparative Models. *J. Asia-Pacific Bus.* 2018, *19*, 286–298, doi:10.1080/10599231.2018.1525250.

88. Cene, E.; Karaman, F. Analysing organic food buyers' perceptions with Bayesian networks: a case study in Turkey. *J. Appl. Stat.* 2015, *42*, 1572–1590, doi: 10.1080/02664763.2014.1001331.

89. Chen, N.H.; Lee, C.H.; Huang, C.T. Why buy organic rice? Genetic algorithm-based fuzzy association mining rules for means-end chain data. *Int. J. Consum. Stud.* 2015, *39*, 692–707, doi:10.1111/ijcs.12210.

90. Fiore, M.; Gallo, C.; Tsoukatos, E.; La Sala, P. Predicting consumer healthy choices regarding type 1 wheat flour. *Br. Food J.* 2017, *119*, 2388–2405, doi:10.1108/BFJ-04-2017-0200.

91. De Sousa Ribeiro, F.; Gong, L.; Caliva, F.; Swainson, M.; Gudmundsson, K.; Yu, M.; Leontidis, G.; Ye, X.; Kollias, S. An end-to-end deep neural architecture for optical character verification and recognition in retail food packaging. In Proceedings of the Proceedings - International Conference on Image Processing, ICIP; IEEE Computer Society; 2018; pp. 2376–2380.

92. Singh, A.; Shukla, N.; Mishra, N. Social media data analytics to improve supply chain management in food industries. *Transp. Res. Part E Logist. Transp. Rev.* 2018, *114*, 398–415, doi:10.1016/j.tre.2017.05.008.

93. Puchalsky, W.; Ribeiro, G.T.; da Veiga, C.P.; Freire, R.Z.; Santos Coelho, L. dos Agribusiness time series forecasting using Wavelet neural networks and metaheuristic optimization: An analysis of the soybean sack price and perishable products demand. *Int. J. Prod. Econ.* 2018, *203*, 174–189, doi:10.1016/j.ijpe.2018.06.010.

94. Zidan, M.; Abdel-Aty, A.H.; El-shafei, M.; Feraig, M.; Al-Sbou, Y.; Eleuch, H.; Abdel-Aty, M. Quantum classification algorithm based on competitive learning neural network and entanglement measure. *Appl. Sci.* 2019, *9*, 1277, doi:10.3390/app9071277.

95. Griepentrog, H.W.; Nørremark, M.; Nielsen, H.; Blackmore, B.S. Seed mapping of sugar beet. In Proceedings of the Precision Agriculture; Springer; 2005; Vol. 6, pp. 157–165.

96. Silwal, A.; Davidson, J.R.; Karkee, M.; Mo, C.; Zhang, Q.; Lewis, K. Design, integration, and field evaluation of a robotic apple harvester. *J. F. Robot.* 2017, *34*, 1140–1159, doi:10.1002/rob.21715.

97. Abdel-Aty, A.H.; Kadry, H.; Zidan, M.; Al-Sbou, Y.; Zanaty, E.A.; Abdel-Aty, M. A quantum classification algorithm for classification incomplete patterns based on entanglement measure. *J. Intell. Fuzzy Syst.* 2020, *38*, 2817–2822, doi:10.3233/JIFS-179566.

98. Onishi, Y.; Yoshida, T.; Kurita, H.; Fukao, T.; Arihara, H.; Iwai, A. An automated fruit harvesting robot by using deep learning. *Robomech J.* 2019, *6*, 1–8, doi:10.1186/s40648-019-0141-2.

99. Makky, M.; Soni, P. Towards Sustainable Green Production: Exploring Automated Grading for Oil Palm Fresh Fruit Bunches (FFB) Using Machine Vision and Spectral Analysis. *Int. J. Adv. Sci. Eng. Inf. Technol.* 2013, *3*, 1, doi:10.18517/ijaseit.3.1.267.

100. Si, Y.; Liu, G.; Feng, J. Location of apples in trees using stereoscopic vision. *Comput. Electron. Agric.* 2015, *112*, 68–74, doi:10.1016/j.compag.2015.01.010.

101. Kondo, N.; Yata, K.; Iida, M.; Shiigi, T.; Monta, M.; Kurita, M.; Omori, H. Development of an end-effector for a tomato cluster harvesting robot. *Eng. Agric. Environ. Food* 2010, *3*, 20–24, doi:10.1016/S1881-8366(10)80007-2.

102. Bulanon, D.M.; Burks, T.F.; Alchanatis, V. A multispectral imaging analysis for enhancing citrus fruit detection. *Environ. Control Biol.* 2010, *48*, 81–91, doi:10.2525/ecb.48.81.

103. Zhao, Y.; Gong, L.; Liu, C.; Huang, Y. Dual-arm robot design and testing for harvesting tomato in greenhouse. *IFAC-PapersOnLine* 2016, *49*, 161–165, doi:10.1016/j.ifacol.2016.10.030.

104. Birrell, S.; Hughes, J.; Cai, J.Y.; Iida, F. A field-tested robotic harvesting system for iceberg lettuce. *J. F. Robot.* 2020, *37*, 225–245, doi:10.1002/rob.21888.

105. Xiang, R.; Jiang, H.; Ying, Y. Recognition of clustered tomatoes based on binocular stereo vision. *Comput. Electron. Agric.* 2014, *106*, 75–90, doi:10.1016/j.compag.2014.05.006.

106. Luo, L.; Tang, Y.; Zou, X.; Ye, M.; Feng, W.; Li, G. Vision-based extraction of spatial information in grape clusters for harvesting robots. *Biosyst. Eng.* 2016, *151*, 90–104, doi:10.1016/j.biosystemseng.2016.08.026.

107. Peng, H.; Zou, X.; Chen, Y.; Yang, L.; Xiong, J.; Chen, Y. Fruit image segmentation based on evolutionary algorithm. *Trans. Chinese Soc. Agric. Eng.* 2014, *30*(18), 294–301.

108. Lee, B.K.; Kam, D.H.; Min, B.R.; Hwa, J.H.; Oh, S.B. A vision servo system for automated harvest of sweet pepper in Korean greenhouse environment. *Appl. Sci.* 2019, *9*, 2395, doi:10.3390/app9122395.
109. Williams, H.A.M.; Jones, M.H.; Nejati, M.; Seabright, M.J.; Bell, J.; Penhall, N.D.; Barnett, J.J.; Duke, M.D.; Scarfe, A.J.; Ahn, H.S.; et al. Robotic kiwifruit harvesting using machine vision, convolutional neural networks, and robotic arms. *Biosyst. Eng.* 2019, *181*, 140–156, doi:10.1016/j.biosystemseng.2019.03.007.
110. Hannan, M.W.; Burks, T.F.; Bulanon, D.M.; (or initial) (or initial) A real-time machine vision algorithm for robotic citrus harvesting. In Proceedings of the 2007 American Society of Agricultural and Biological Engineers: St. Joseph, MI, Minneapolis, MA, June 17–20, 2007; 2007; Vol. 8 BOOK, pp. 1–11.
111. Mureşan, H.; Oltean, M. Fruit recognition from images using deep learning. *Acta Univ. Sapientiae, Inf.* 2017, *10*(1), 26–42.
112. Arefi, A.; Motlagh, A.M.; Mollazade, K.; Farrokhi Teimourlou, R. Recognition and localization of ripen tomato based on machine vision. *Aust. J. Crop Sci.* 2011, *5*(10), 1144–1149.

5

IoT-Enabled Services for Sustainable Municipal Solid Waste Management in India

Hrishikesh Chandra Gautam, Vinay Yadav, and Vipin Singh

CONTENTS

5.1 Introduction

Rapid urbanization and industrial development have led to an increase in material consumption due to affluent lifestyles, and as a result increases in per capita municipal solid waste (MSW) have been generated (Yadav et al., 2016, 2020). The increase in the amount of solid waste generated has turned efficient and sustainable management of solid waste into a challenge. The complex system of solid waste management (SWM), which comprises efficient collection and segregation to proper disposal, reuse, and recycling of the waste generated, is facing challenges due to increases in population and solid waste quantities. These challenges are faced by officials from sanitation departments, city municipalities, decision makers, as well as ordinary citizens on a daily basis. A lot of studies have been done in recent decades in various fields, ranging from efficient collection and disposal to the economic sustainability of the collection process (Yadav et al., 2021). In the studies, the SWM infrastructure in developing countries like India is found to be improper and inadequate compared to developed countries in the West. The problem is compounded by the momentous growth in population and affluent lifestyle of the population.

5.2 Municipal Solid Waste Management in India

MSW Management is a challenging task due to the huge amount (85 million tons) of municipal solid waste per year by 377 million people living in 7,935 towns and cities. Of MSW, 50% of waste is collected, 14% treated, and 36% reaches landfill sites (Samar, 2019). Per capita MSW generated in developing countries of around 0.3–0.5 kg per person per day is smaller compared to industrialized countries of 0.8–1.4 kg per person per day, but due to lack of infrastructure and monetary resources it poses huge problems. According to CPCB (2016), approximately 62 million metric tons (MT) MSW was generated yearly in India, which is ~0.45 kg per capita per day. This is expected to increase by 165 MT, 230 MT, and 436 MT by 2030, 2041, and 2050, respectively. About 82% of generated MWS is being collected, out of which only 28% is treated, the remaining being openly dumped (Sharma and Jain, 2019). The revenue allotted to SWM does not compensate the monetary requirement for minimization and treatment of solid waste. As a cheaper option, most of the generated waste finally goes to landfill sites, but due to limitations of space required and the increasing population the option is not sustainable for a long duration. Apart from that, a lot of waste generated in developed countries goes to landfill sites in Asian and African countries as a cheap solution, which increases the burden on landfill sites in developing and poor countries. This issue of Not In My Backyard (NIMBY) has been widely reported in literature and increases the severity of an already dire situation (Sonak et al., 2008; Guerrero et al., 2013).

In recent years the rise of information technology, artificial intelligence (AI), machine learning (ML), and Internet of Things (IoT) in general has made MSW management efficient on many fronts (Atzori et al., 2010). With the help of faster computation, involvement of big data, better maintenance of infrastructure through efficient data handling, and understanding the core of problems through data analytics, better handling and management can be achieved.

5.3 Internet of Things (IoT)

With the rapid rise in technology and digitalization over recent years, a lot of devices that we commonly use in our lives as well as industries have been replaced by so-called smart devices which possess microprocessors for faster and more efficient utilization. These devices are normally connected to surrounding devices through Bluetooth or Wi-Fi, where data can be shared over a network of devices and can enhance the performance of a device at an individual as well as network level. This enhancement of performance and connectivity lead us to the concept of IoT.

IoT can be defined as interconnecting physical devices like sensors and actuators into wired or wireless networks for the achievement of specific tasks (Alqahtani et al., 2019; Saha et al., 2017; Thakker et al., 2015). The interconnectivity helps the device to share the data and information with others across the platform by a unified network which can enhance the efficiency of the device and perform a variety of actions to develop innovative products and solutions. The basic idea of IoT is to connect and share information between radio frequency identification tags, sensors, actuators mobile phones, and smart devices online or through ad hoc networks so that the devices can operate more efficiently and at a larger scale (Arasteh et al., 2016; Kim et al., 2017; Anagnostopoulos et al., 2015; Kunst et al., 2018; Shyam et al., 2017). Recent projects have also been involved in applications in smart city development as well as industrial IoT. Even though the applications are at a nascent stage, their potential in making our everyday life easier and more efficient is huge. Areas where these technologies are being explored and research is being conducted are solid waste management, health, supply chain management, connecting houses, buildings, and cars, etc.

The issues that have to be resolved for efficient application in all these areas include better analytic and monitoring tools for the network, efficient and fast updating of data storage, as well as backup, security issues involving data access and data theft, and fast internet access to all the devices at different times. A lot of recent applications are paying attention to these issues, and rapid development is taking place in resolution of these issues.

5.4 IoT Applications in MSW Management

In the case of MSW, the IoT solution has been applied to a number of steps including waste collection, transportation of waste, waste segregation, waste recycling etc. (Weber et al., 2017; Zanella et al., 2014). Application of IoT in efficient waste collection involves detecting the filling level, tilt, fume and gas generation, GPS location, humidity, temperature sensing, and detection of fire. If any of these parameters are not found within the desired value or limit, the nearest person with the responsibility of MSW collection will be prompted to take the desired steps in alleviating the issue. These parameters have been considered in development of smart waste bins which have been developed and are operational in some municipalities (Islam et al., 2012; Mahajan et al., 2014; Sinha et al., 2015).

In the case of waste transportation/collection, the collection vehicle has to traverse through the city to collect MSW disposed in bins. The IoT system detects through sensors when a specific waste bin is filled more than a given level and send a signal to the driver of the collection truck. The driver gets the location of the bins which have been filled more than the desired level and have to be emptied. A routing algorithm is applied to the location of bins and the route traversed for the waste transportation with the least distance and fuel expenditure, keeping in mind the traffic across the area travelled (Arora et al., 2020; Varsha et al., 2019a, 2019b). Sensors also take account of the amount of waste filled in the truck and the distance from the dumping yard, keeping in mind the route, time, and duration. Sensors can also detect the emission of flues/poisonous gases in the collected waste (Figure 5.1).

Nidhya et al. (2020) developed an enhanced route election algorithm to decide the path of the waste collection truck so that the truck can traverse through the city and collect waste from smart bins filled more than 90% using the ERS algorithm, using the shortest route thereby saving time, fuel, and producing low emissions. In the study, they proposed a system comprising smart bins, remote servers, and base station. The smart bin is equipped with a sensor which sends a signal to the remote server through the base

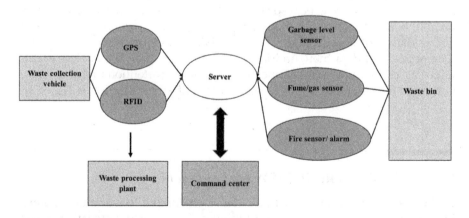

FIGURE 5.1
An outline of a framework for IoT applications in the collection of processing of waste under the solid waste management plan.

station when the smart bin is filled more than 90%. The remote server is connected to all the smart bins maintained by the city corporation. The base station is used for referencing the smart bin and its geolocation.

The ERS algorithm first directs the collection vehicle to the closest filled smart bin and then directs the vehicle to the next smart bin with the shortest travel time. Otherwise, the nearest smart bin is selected until the end of the collection procedure. The algorithm and the collection system developed takes care of collection and monitoring of garbage in real time, complexity of route detection in pickup of garbage in multiple locations, as well as end-to-end delays in data transmission between the smart bin and the remote server.

Murugesan et al. (2019) proposed a framework-based model on waste level detection in waste bins. The data, generated from the level of waste and detected with sensors, is transmitted to officials through the internet. The data is also used to detect unwanted waste bins which can be removed or transferred to other locations for enhancing the route and structure of the MSW collection network. The data is also used to understand the distribution and trends of the waste collection status and distribution to allocate the collection manpower and resources more efficiently in the future. The spatial analysis helps in detecting unwanted waste bins and also to anticipate the waste collected status of waste bins area-wide. A sensor hub comprises a bridge rectifier, step-down transformer, a channel of circuit, and a device to regulate voltage, with data transfer through an ethernet modem connected to Arduino UNO microcomputer board. Ultrasonic sensors are used to detect the level to which the waste bin is filled, and the data is communicated to the nearest control room through an HTML based webpage. The ultrasonic sensor uses sound waves to detect the height of waste collected in the waste bin. Downpour sensors are used to detect precipitation and IR sensors are used to detect proximity with objects kept near the waste bin. In case of precipitation, the waste bin lid is automatically locked with the help of motors. Another study by Malapur et al. (2017) proposed an MSW management system to provide an optimized path for waste collection vehicles using dynamic scheduling. The waste bins provide alerts when filled up using sensors. A user-friendly android app helps in the optimizing collection of waste. Nirde et al. (2017) proposed an IoT-based wireless solid waste management system for smart cities which helps municipal bodies with continuous monitoring of the waste level in waste bins remotely, using a web server, thereby saving time and optimizing costs. The authorities get informed of filled waste bins through a message-using Global System for Mobile Communications (GSM) placed in the waste bin. Garbage collection vehicles are sent to the relevant location for waste collection. Poddar et al. proposed an integrated system for waste management, using smart waste bins equipped with a network of sensors. The system also transmits real-time data indicating the waste level of the bin (Poddar et al., 2017).

Kumar et al. (2016) proposed an IoT-based alert system for waste collection, which sends an alert to the municipal web server, based on the garbage level in dustbins, to empty the dustbins with proper verification. The system is supported by a module integrated with RFID and IoT. Baby et al. (2017) proposed a waste alert system that alerts the municipality to collect waste from filled up waste bins. The garbage trucks are sent to only those areas with filled garbage bins, saving time and decreasing fuel use and vehicular emissions. The collected data is used to train machine-learning based models to get an understanding of waste generation trends. The model results are used to predict the waste bins that are going to be filled soon. Pardini et al. (2018) proposed a smart waste bin with load cell sensors and ultrasonic sensors used for identification, Global Positioning System (GPS) for determining the location and Global System for Mobile Communications (GSM) or General Packet Radio Service (GPRS) for communication. The study intends to contribute to social, economic, and environmental management of large cities.

Even with flexibility in structure of IoT devices, sensors, and sensor networks, there are issues related to speed in data collection, quality of data collected, as well as connectivity across the network and data privacy (Gubbi et al., 2017). To address these issues, standard structures consisting of various layers have been proposed over the years. These structures are proposed with the objective of efficient quality of standards (QoS), sustainability, data integrity, confidentiality, and reliability.

5.5 Structure of IoT Framework

The different components and layers in the IoT framework are described below. Even though the detailed structure can vary based on application, the overall framework follows the following structure (see Figure 5.2).

5.5.1 Perception Layer

Perception layer of IoT architecture is also defined as the physical layer of the network as it constitutes the hardware allocated with the responsibility of collecting data in the form of physical information, processing the data, and transmitting it to the higher layer securely. It uses the application of sensors to detect the physical information from the surrounding such as weight temperature, humidity etc. In addition, the data can be collected through QR codes and RFID tags. In the case of solid waste management systems, the perception layer consists of sensors which collect the data regarding level of waste collected, weight, humidity, presence of gas and fumes, temperature, detection of fire, precipitation, as well as nearby objects, and sends the information to the higher layers.

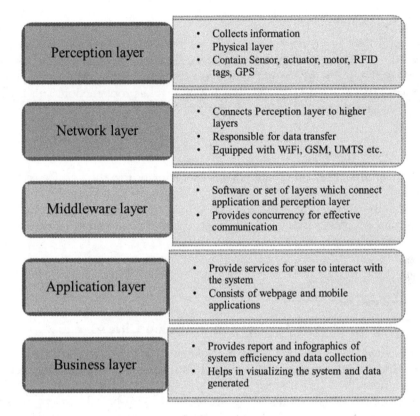

FIGURE 5.2
Detailed architecture of IoT infrastructure.

5.5.2 Network Layer

The network layer is responsible for collection of data from the perception layer and transferring the data to higher layers where the processing system for the collected data is located.

The layer uses a single or combination of different connectivity protocols like GSM, UMTS, Wi-Fi, Infrared etc. In addition to connecting the different components of the IoT network, the layer also has the responsibility to perform cloud computing tasks and overall data management.

5.5.3 Middleware Layer

This is the layer consisting of software or a set of layers used to interconnect the components which are unable to communicate otherwise. The main objective of this layer is to provide concurrency between the perception layer and application layer so that they can interact effectively and efficiently, and it also plays in modification and development of new IoT infrastructures.

5.5.4 Application Layer

This layer does not contribute to the overall structure of the IoT infrastructure directly, but it provides the various services and platforms for the users to interact with the IoT system and access the information produced by the infrastructure and interpret the information to take proper action. This layer may consist of webpages, Android apps etc. or a combination of them.

5.5.5 Business Layer

This layer manages the overall IoT system including the service-related applications and reports. This layer is responsible for providing the analysis report of the underlying layers and over-efficiency of the IoT application. The layer also addresses issues related to connectivity, data speed, processing time, and privacy.

5.6 IoT-Based SWM Application in Indian Cities

IoT based solutions for solid waste management have been applied to a number of Indian cities under the Swachh Bharat Abhiyaan Programme, operational under the Ministry of Housing and Urban Affairs (MoHUA, 2019). The solution ranges from smart waste bins using IoT-based sensors to use of RFID tags for waste collection vehicles and automated weighbridges for weighing the waste collected.

5.6.1 Bengaluru

The city of Bengaluru generates a large amount of waste amounting to 4,500 tons per day. The collection and transportation of waste from different locations in the city require 4,000+ primary collection vehicles (PCV) and 500+ secondary transportation vehicles (STV). IoT-based technologies have been applied for monitoring and regularizing of the fleet movement of collection vehicles to the destinations for waste collection and recording of the tonnage of waste delivered at designated waste collection yards as well as landfills. All the waste collection vehicles (PCVs and STVs) under Bruhat Bengaluru Mahanagara Palike (BBMP) are installed with RFID tags.

Installation of RFID tags ensures that only authorized vehicles are allowed at the designated destinations and unauthorized transfer of waste can be stopped. The RFID applications are also used to record the PCV's/STV's weight at the weighbridge as an essential data field to the scanning process. This daily data is used for calculating payments to be given to service providers based on the waste collection vehicle's performance. The details of all the operational waste collection vehicles are uploaded to the Auto Tipper Registration (ATR) application by the official in charge of the waste collection. The PCVs are provided with RFID tags which are fixed on the vehicle for easy and fast scanning. The RFIDs of all the PCVs are scanned at the mustering point as well as the first and second transfer points to the STVs. The regular scanning of vehicles helps in recording vehicle attendance as well as completion of the required number of trips for the day and transfer of waste to the designated STVs. The RFID scanning application is installed in the presence of authorized personnel with their approval and authentication. Every vehicle provided with an RFID tag has to be scanned while entering the collection/processing plants or sanitary landfill. The vehicle is allowed access after authorization through the RFID tags provided to the waste collection vehicles. The data provided through RFID tags of different vehicles is compiled and sent to the cellphones of the SWM officials on a daily basis. The RFID-based monitoring with centralized control room is integrated into the blockchain-based citizen helpline. The daily data of vehicle movement through designated destinations and vehicle performance is analyzed at the end of the month.

The RFID-based smart monitoring system monitors and analyses the movement as well as performance of all the vehicles in a fast and efficient manner. All the vehicles reach the designated destination; and there is no possibility of data manipulation with regard to vehicle performance. The elimination of data manipulation makes the collected data more reliable and authentic. It enables the officials to analyze the number of vehicles that arrive against the space allocated for vehicles as well as the input of waste reaching the processing plants and the sanitary landfills. This helps in optimization of vehicles, waste bins, and manpower required based on the amount of waste transferred and total number of trips made to collect the waste. The integrated system has improved the overall efficiency of waste collection and transportation in a seamless manner and increased the overall quality of work for the designated officials.

5.6.2 Vijaywada

Vijaywada is a city in Andhra Pradesh state in India with the waste generation of 550 metric tons/day. All the solid waste management bins are installed with RFID tags for monitoring the waste disposal and collection process and increasing the efficiency and speed of the overall process. The RFID tags are read in a timely manner by the RFID readers, and the collection of waste from the bins with the help of the waste collection vehicle is recorded. The movement of the waste collection vehicle between the waste bins and waste processing plant is recorded in real time using a GPS system. All the data collected through the IoT-based system is transferred to the central command center where the data is processed and evaluated to analyze the overall efficiency of the system, and corrective measures are taken to enhance the efficiency. The entire process of waste collection and transportation to processing sites and landfill is monitored through a structured process using IoT-enabled devices and networks. RFID tags installed on top of each waste bin are allotted a tag with unique details (serial number, location, collection vehicle details etc.). Once the collection vehicle reaches the location, the driver can read the RFID tag with his RFID reader, and the information is sent to the server with waste bin details and time of lifting, and the database is updated. SIM-based solar close circuit TV (CCTV) cameras which require low maintenance are installed across the city to monitor the condition of waste bin as well as spillage of waste. The vehicular movement is recorded using GPS-based vehicle tracking devices and updated to the server in real time for online monitoring. Timely lifting of garbage using IoT-based garbage monitoring saves manpower and day-to-day operating costs, and the frequency of complaints is also reduced.

5.6.3 Surat

In Surat, a city in Gujrat state, IoT applications were developed to ensure that the collection vehicles attend the specified route at the defined time for collecting waste. The system was also used for measuring the performance of vehicle/contractor and calculation of payment and penalty based on the performance. The system is able to generate reports of the vehicle and waste collection performance on demand.

The system monitors the vehicle in real time providing the information of waste collection to the server. The transfer and disposal sites are also automated to increase efficiency and also minimize human intervention. The door-to-door vehicles are installed with radio frequency identification (RFID) tags that identify the vehicle in real time at the transfer stations and record the weight of waste collected which is updated on the server.

The system tracks a total of 551 vehicles where GPS is used for real-time tracking of location to check whether the door-to-door vehicle has traversed through all the points of interest and collected waste along the route assigned for the specific vehicle. The contractors can also be penalized based on the number of points of interest not covered along the route. GPS devices provide real-time monitoring information of all the door-to-door collection vehicles to the command-and-control center of Surat Municipal Corporation. Vehicle information and the points of interest traversed can also be linked to the weight of collected waste automatically recorded at the weighbridge.

The system works with minimum human intervention to provide accurate real-time data for each vehicle at all the waste handling facilities. The system is accountable and transparent which prevents misuse by avoiding manual intervention. It ensures real-time coverage of all of Surat city with actionable data for efficient decision making. The software application part of the system is also being used for redressal of public grievances.

5.6.4 Vapi

An IoT-based system for waste collection, disposal, and addressing user complaints in real time was developed for the city of Vapi. The system is currently being used by Vapi Municipal Corporation.

The system used near field communication (NFC) tags that were provided to every house. It solves the problem of difficulty in finding addresses by municipal employees. It also helps the municipal corporation to integrate all government services into individual houses.

A digital easy city code is provided to all households using the NFC tag which validates all the visits by door-to-door collection facilities. It also helps the citizen to locate all houses digitally and share the location. Smartphone-enabled complaint management helps to record all citizens grievances in real-time and provide an efficient way for waste management officials to resolve issues in a fast and efficient manner. Citizens connect using SMS, and phone call alerts are also sent to citizens to provide updated information. Easy city code is an open smart address system which can be used by citizens to access the geolocation of a given address and is easy to find and share.

The developed system has reduced the grievances received related to door-to-door collection of waste by 90%, and the number of grievances solved per day has increased ten-fold. The system helped the waste department save 20+ lakh rupees a day and reduced the time taken in addressing and solving grievances from two days to one day. The system has also added accountability and transparency regarding the working of Vapi Municipal Corporation. Timely data-oriented reports and analytics also help in improving the system and efficient decision making.

5.6.5 Bhopal

Bhopal, a city in the state of Madhya Pradesh, was facing the issue of inefficient waste management due to its wide geographical area, with bins placed at different locations and which fill at different rates i.e., bins in areas with high traffic rates and population are filled at a faster rate. Aligning the garbage collection trip with the status of filled garbage bins was not possible, leading to a higher number of trips with excess fuel expenditure and operating costs.

The municipal corporation has installed 700 RFID tags and fuel sensors in waste collection vehicles as well as 230 IoT-based sensors in 460 twin bins across high priority zones identified by the authorities. A real-time vehicle tracking system (RTVTS) was integrated with real-time monitoring using IoT sensors for seamless and efficient collection of waste across the city. The integration helped in route planning as well as optimizing the waste collection, resulting in a high frequency of weekly waste collection and a reduction in grievances by citizens. The twin bins in the identified zones were installed with IoT-based ultrasonic sensor devices which identify the fill status and alert the central command center in real time once the waste bin reaches 80% of its fill capacity. Accordingly, the nearest vehicle enroute is identified and instructed through SMS triggered by the RTVTS system to collect waste from the identified garbage bin. The integrated system helps in route planning of the garbage collection by vehicle as well as ward/area-wise route optimization in the city for effective mapping of the vehicle, route planning, and improved monitoring of waste collection. Real-time data analysis helps in comparing the ward/area-wise data collection process of the city i.e. spatially as well as temporally. It helps in efficient decision making and avoiding unnecessary trips. It also helps in weekly analysis of waste collection trips, creating reports on fuel efficiency, cost, and manpower optimization.

5.6.6 Indore

Indore city in Madhya Pradesh has a waste generation capacity of 1,100 tons per day. The city's municipal corporation was facing the issues of not knowing whether all the waste had been collected, assessment of the quantity of wet and dry waste collected, as well as data manipulation in the amount of waste collected and trips taken. An integrated IoT-based system was developed for the city to address these issues in a timely and efficient manner. The door-to-door collection vehicles were equipped with GPS and RFID tags. The vehicles are automatically read with the help of an RFID tag at the respective transfer station, and entry of unauthorized vehicles is stopped. Real-time data is recorded, transferred, and analyzed at the command center. When the waste collection vehicle equipped with GPS and RFID reaches the weighbridge, the automated barrier can read the RFID tag and open. The vehicle weighing operation is conducted in three stages. At the first weighbridge the vehicle is weighed. Then the vehicle unloads the dry waste, and weight is recorded at the second weighbridge. Wet waste is unloaded, and the vehicle is weighed at the third weighbridge. The compiled data is communicated to the central command center in real time. The data of distance travelled between collection sites and the processing facility, and the amount of dry and wet waste collected by each vehicle is recorded and analyzed for all departments at the command center. The system has optimized the day-to-day operation costs, time spent, and manpower utilized. The system has stopped the manipulation and tampering of collected data by removing human intervention in the data collection process. The accurate data is used for future planning and development as well as further optimization of the waste collection process.

5.7 Conclusions

The current work deals with the application of IoT in solid waste management in Indian cities. The current situation of solid waste management in

India was described, followed by the description of IoT-based systems and their different applications. The chapter ends with a detailed account of IoT-based systems applied in different cities across India. Various benefits and shortcomings of the technology were discussed as summarized below.

The benefits of application of IoT in developing a smart SWM are seamless and allow real-time connectivity and flow of information. It helps the stakeholders (citizens as well as municipal organization officials) to connect and share information regarding waste collection and disposal through updates on smartphone applications or SMS. IoT makes the SWM system more efficient in terms of waste collected, time spent, as well as money spent on vehicle fuel.

The system also minimizes human intervention which thereby reduces hindrances caused by unintentional human error as well as data manipulation. The data is recorded and stored at a central server from which analytic reports are generated on efficiency and resources utilized, and these communicated to shareholders. Data trends are also used for future improvements in the system.

The system also has some shortcomings such as investment in installation being required, as well as maintenance. Even though the system is developed with devices which need low maintenance, technical knowledge and support will be required in case of maintenance as well as handling of errors and faults.

The public as well as employees at waste collection facilities must be educated on using the system efficiently through mobile applications. Initial training may be required for the officials to access the data and make sense of the system.

Lack of accessibility of fast internet and mobile communication at the location can also act as a roadblock in the application of IoT-based systems in developing an efficient solid waste management system.

Machine learning-based applications in combination with big data systems can be used to gain knowledge about the waste generation. Waste generation can be forecasted, and based on the results of this, waste handling officials can be informed in advance. A combination of spatial interpolation and clustering techniques can also help in identifying the hotspots where extra resources could be allocated. Further research is required to reduce the overall costs and provide better data connectivity in remote locations.

References

Alqahtani, A. Y., Gupta, S. M. and Nakashima, K. 2019. Warranty and maintenance analysis of sensor embedded products using internet of things in industry 4.0. *International Journal of Production Economics*, 208, 483–499.

Anagnostopoulos, T. Zaslavsky, A. and Medvedev, A. 2015. Robust waste collection exploiting cost efficiency of IoT potentiality in Smart Cities. In: Proceeding of 1st IEEE International Conference on Recent Advances in Internet of Things, RIoT 2015, Singapore.

Arasteh, H., Hosseinnezhad, V., Loia, V., Tommasetti, A., Troisi, O., Shafie-khah, M. and Siano, P. 2016. IoT-based smart cities: a survey. In: IEEE 16th International Conference on Environment and Electrical Engineering (EEEIC), pp. 1–6.

Arora, G., Kumar, A. Versha and Kumar, N. 2020. Swarm Intelligence based QoS optimized routing in WSN. *Test Engineering and Management*, 82, 12880–12885.

Atzori, L., Iera, A. and Morabito, G. 2010. The Internet of Things: a survey. *Computer Networks*, 54, 2787–2805.

Baby, C.J., Singh, H., Srivastava, A., Dhawan, R. and Mahalakshmi, P. 2017. Smart bin: an intelligent waste alert and prediction system using machine learning approach. In: International Conference on Wireless Communications, Signal Processing and Networking (WiSPNET), Chennai, pp. 771–774, doi: 10. 1109/ WiSPNET.2017.8299865.

CPCB. 2016. *Central pollution controlboard (CPCB) bulletin*. Goverment of India. Available at http://cpcb.nic.in/openpdffile.php?id=TGF0ZXN0RmlsZS9MYXRlR c3RfMTIz X1NVTU1BUllfQk9PS19 GUy5wZGY.

Gubbi, J., Buyya, R., Marusic, S. and Palaniswami, M. 2017. Internet of Things (IoT): a vision, architectural elements, and future directions. *Future Generation Computer Systems*, 29(7), 1645–1660.

Guerrero, L.A., Maas, G. and Hogland, W. 2013. Solid waste management challenges for cities in developing countries. *Waste Management*, 33(1), 220–232. doi: 10.1016/j.wasman.2012.09.008.

Islam, M.S., Hannan, M.A., Arebey, M. and Basri, H. 2012. An overview for solid waste bin monitoring system. *Journal of Applied Sciences Research*, 5(4), February 2012. ISSN 181-544X.

Kim, T.H., Ramos, C. and Mohammed, S. 2017. Smart city and IoT. *Future Generation Computer Systems*, 15, 9–162.

Kumar, N.S., Vuayalakshmi, B., Prarthana, R.J. and Shankar, A. 2016. IoT based smart garbage alert system using Arduino UNO. In: IEEE Region 10 Conference (TENCON), Singapore, pp. 1028–1034, doi: 10.1109/TENCON.2016.7848162.

Kunst, R., Avila, L., Pignaton, E., Bampi, S. and Rochol, J. 2018. Improving network resources allocation in smart cities video surveillance. *Computer Networks*, 134, 228–244.

Mahajan, K. and Chitode, J.S. 2014. Waste bin monitoring system using integrated technologies. *International Journal of Innovative Research in Science, Engineering and Technology*, 3(7), July 2014. (An ISO 3297: 2007 Certified Organization).

Malapur, B.S. and Pattanshetti V.R. 2017. IoT based waste management: an application to smart city. In: International Conference on Energy, Communication, Data Analytics and Soft Computing (ICECDS), Chennai, pp. 2476–2486, doi: 10.1109/ICECDS.2017.83898 97.

MoHUA. 2019. *Transforming urban landscapes of India Success Stories in Information & Communications Technology (ICT)*. Report by Swachh Bharat Mission (MoHUA).

Murugesan, H., Revathy, N. and Senthil Kumar M. 2019. IoT-enabled service management system for smart dining environment IoT-enabled service management system for smart dining environment. *Journal of Advanced Research in Dynamical and Control Systems, Institute of Advanced Scientific Research, Inc.*, 11(3), 1823–1828.

Nidhya, R., Kumar, M., Ravi, R.V. and Deepak, V. 2020. Enhanced Route Selection (ERS) algorithm for IoT enabled smart waste management system. *Environmental Technology & Innovation*, 20, 101116.

Nirde, K., Mulay, P.S. and Chaskar, U.M. 2017. IoT based solid waste management system for smart city. In: International Conference on Intelligent Computing and Control Systems (ICICCS), Madurai, pp. 666–669, doi: 10.1109/ICCONS.2017.8250546.

Pardini, K., Rodrigues, J.J.P.C., Hassan, S.A., Kumar, N. and Furtado, V. 2018. Smart waste bin: a new approach for waste management in large urban centers. In: 88th Vehicular Technology Conference (VTC-Fall), Chicago, IL, pp. 1–8, doi: 10.1109/VTCFall.2018.8690984.

Poddar, H., Paul, R., Mukherjee, S. and Bhattacharyya, B. 2017. Design of smart bin for smarter cities. In: Innovations in Power and Advanced Computing Technologies (i-PACT), Vellore, pp. 1–6, doi: 10.1109/ICCONS.2017.8250546.

Samar, L. 2019. India's challenges in waste management. *Down To Earth.* 08 May https://www.downtoearth.org.in/blog/waste/india- s- challenges- in- waste- management- 56753.

Saha, H. N., Auddy, S. P., S. Kumar, S. Pandey, S. Singh, R. and Saha, S. 2017. Waste management using Internet of Things (IoT). In: 8th Annual Industrial Automation and Electromechanical Engineering Conference (IEMECON), pp. 359–363. IEEE.

Sharma, K. D. and Jain, S. 2019. Overview of municipal solid waste generation, composition, and management in India. *Journal of Environmental Engineering*, 145(3), 04018143-1-18.

Shyam, G.K., Manvi, S.S. and Bharti, P. 2017. Smart waste management using Internet-of-Things (IoT). In: 2017 2nd International Conference on Computing and Communications Technologies, ICCCT, pp. 199–203.

Sinha, T., Kumar, K.M. and Saisharan, P. 2015. Smart dustbin. *International Journal of Industrial Electronics and Electrical Engineering*, 3(5), 101–104, May 2015. ISSN: 2347-6982.

Sonak, S., Sonak, M. and Giriyan, A. 2008. Shipping hazardous waste: implications for economically developing countries. *International Environmental Agreements: Politics, Law and Economics*, 8(2), 143–159. doi: 10.1007/s10784-008-9069-3

Thakker, S. and Narayanamoorthi, R. 2015. Smart and wireless waste management. In 2015 International Conference on Innovations in Information, Embedded and Communication Systems (ICIIECS), pp. 1–4. IEEE.

Varsha, M. B., Kumar, M. and Kumar, N. 2019a. Development of QoS optimized routing using Artificial bee colony and TABU-GA with a mobile base station in Wireless Sensor Network. *International Journal of Innovative Technology and Exploring Engineering (IJITEE)*, 9(1), 926–933, November 2019. ISSN: 2278-3075.

Varsha, M. B., Kumar, M. and Kumar, N. 2019b. Hybrid TABU-GA search for energy efficient routing in WSN. *International Journal of Recent Technology and Engineering (IJRTE)*, 8(4), November 2019, 3250–3256. ISSN: 2277-3878.

Weber, M., Lucic, D. and Lovrek, I. 2017. Internet of Things context of the smart city. In International Conference on Smart Systems and Technologies (SST) IEEE Conference Publications 2017, 187–193.

Yadav, V. and Karmakar, S. 2021. Multi criteria decision making for sustainable municipal solid waste management systems. In: *Computational Management* (pp. 587–598). Springer, Cham.

Yadav, V., Karmakar, S., Dikshit, A. K. and Vanjari, S. 2016. Transfer stations siting in India: a feasibility demonstration. *Waste Management*, 47, 1–4.

Yadav, V., Kalbar, P. P., Karmakar, S. and Dikshit, A. K. 2020. A two-stage multi-attribute decision-making model for selecting appropriate locations of waste transfer stations in urban centers. *Waste Management*, 114, 80–88.

Zanella, A. and Vangelista, L. 2014. Internet of Things for Smart Cities. *IEEE Internet Things Journal*, 1, 22–32.

6

IoT-Enabled Smart Dustbin

Sankalp Nayak, Alok Narayan, and Jitendra Kumar Rout

CONTENTS

6.1 Introduction

As far as the present-day scenario goes, the primary cause of land pollution is waste overflow which may accumulate and lead to severe threats on humankind and all living things at large. Therefore, the project of "smart connected dustbins" was designed. In the first place, it tackles waste disposal and, in due course, prevents the problems caused otherwise. The project uses concepts of IoT, Android applications, and cloud technologies to deploy the idea as mentioned above.

The smart dustbin works based on a variety of sensors. The IoT concept is widely used as it helps to connect various components over the internet and aids in intercommunication as and when required. Primarily sensors are located all over the bins to track their locations, as they are spread across different places and have to be coordinated from a single site. A unique ID is provided for each bin to aid in proper identification. Various other sensors are used like ultrasonic sensors, which track the level of garbage in the bins, the master Arduino catcher with Wi-Fi module which sends an alarm to relevant authorities when the threshold limit of the bin is reached, and then the mobile applications or message services can be used for taking a quick view. Further, another detector, i.e., the odor detector, can be used to

DOI: 10.1201/9781003184096-6

notify officials about the rotting smell of the garbage so that harmless chemical sprinklers can be spread. The tracking of the bins and their monitoring becomes very viable using mobile applications.

The research aims at designing and building a model to estimate the amount of content inside the smart dustbins. The authorities are sent an app notification when the content of the bin exceeds 80%. The tracking can be easily maintained with the help of the unique ID, and the collected data can be easily viewed on the back end with the help of a mobile app or dashboard.

a. **Problem definition:** Domestic garbage is disposed of in dustbins positioned along the roadway. Every public garbage bin is emptied at random. Because the dustbins occasionally fill up quicker than usual, constant human supervision is necessary to manage the overflow level. People cannot throw their garbage in dustbins when it has overflowed so they dump it outside the bin. During the rainy season, the situation deteriorates as rainwater enters the trash, creating a foul odor. And as we have seen all around us, dustbins are overflowing, and the involved garbage collecting authorities frequently do not get notification within the time frame specified. Moreover, the garbage collection trucks don't know which dustbin needs to be emptied; instead, they sequentially visit dustbin sites where many dustbins don't need to be emptied as they might not be full. The process of visiting sequentially where some dustbin might not be required to be emptied consumes a lot of time and energy, such as wastage of fuel. It reduces the overall efficiency of the waste collection system.

b. **Need for improvement in waste management system**
    i. One of the most crucial things that every human being values is cleanliness. We must acknowledge filthy places and make them tidy and nice in order to be clean. This is also beneficial to the tranquility of the environment. Dustbins are generally seen as dirty since they are overflowing with trash and emit terrible odors.
    ii. Waste management reduces the volume and hazardous character of residential and industrial waste, preventing harm to the inhabitants and the environment as well.
    iii. By the end of 2030, at least two-thirds of the entire population across the globe will live in cities. For urban areas, this type of waste management system is essential.

c. **Features of the smart dustbin**
    i. It emphasizes the concept of "digital India."
    ii. It can check the form of the waste (i.e., solid or liquid waste).
    iii. The smart dustbin will check the amount of waste present in it and will periodically send the messages to the appropriate authority directly.

    iv. The smart dustbin app will also notify the municipal corpora-
tion workers about the shortest path available to the dustbins
using "transportation software."

d. **Advantages of the smart dustbin**

    i. The Global Positioning System (GPS) shows the shortest path
available, saving time and fuel to a great extent.

    ii. The system is used to indicate the amount of waste that has been
disposed of.

    iii. Webpages make it simple to keep track of the dustbin.

    iv. Daily seasonal information is gathered from the e-waste man-
agement. Cleaning companies can plan efficiently and send their
staff to empty the trash and the routes their cleaners should take
for the least amount of trip.

    v. Our smart operating system enables communication between
dustbins and service operators

    vi. The sensors of the containers provide real-time information that
will lead to the ability to prioritize collection processes.

    vii. Reduces the infrastructure load and the operating and mainte-
nance costs of the service by up to 30%.

    viii. Reduces the amount of time spent by human monitoring. The
system may be used to reduce costs and time.

e. **Applications of the smart dustbin**

    i. Paves the way for digitalization of the entire globe and smart
connected cities.

    ii. It reduces time and increases the efficiency of waste control and
management.

The rest of the chapter is organized as follows: details of the related works
are discussed in Section 6.2. The details of the proposed IoT-enabled smart
dustbins are given in Section 6.3. Section 6.4 depicts the proposed architectural details. Finally, Section 6.5 concludes the chapter.

6.2 Related Work

Upon a survey all over India's street disposal bins, it has been found that due
to lack of monitoring, there have been cases that have led to overflowing conditions of the dustbins that contributed to the pollution of neighboring places
and led to many diseases. Obviously, the concerned authorities can't keep

track manually of the various sites those bins are located. Hence as a boon to this problem, we have come up with an innovative approach of the "smart connected dustbins", which narrow down the management problem to a great extent and thus lead to the disposal of this harmful waste more efficiently.

The existing projects have been of great help for our project even though they are not as efficient as the ones we have developed. IoT concepts have been implemented for establishing an interconnection between the various bins spread across the place, and the use of Android applications has been the prime source of proper monitoring and management of this entire process.

Parikh et al. [1] have given an abstract idea about smart dustbins, which are basically built upon components like ultrasonic sensors, servo motors, and microcontrollers. On sensing the person's presence automatically, the dustbin accepts the RFID from the user and displays the person's name on the serial LCD located at the front of the dustbin. The servo motor opens the dustbin door until the person leaves the place from the time his presence is detected initially, which is detected by the ultrasonic sensors. After the entire thing is over, the GSM attached to the dustbin sends payback points to the users, which they can redeem using an Android application "DUSTBIN". If the dustbin is full, it automatically sends a message to the control room by turning on the red LED attached to it. In a nutshell, this dustbin helps clean India.

A similar work by Damakale et al. [2] proposed an idea that mainly focuses on providing a reliable and easy solution to garbage disposal. This extends to all types of users. This project basically tackles the problem of GMS by introducing IoT, which works on the basis of the amount of garbage present in the dustbin. This modernized system of bringing IoT into the foreground becomes a real-time project. Also, the further applications of system MP Labs and CSS compiler to program microcontrollers make it an even more efficient system.

Mamun et al. [3] have introduced a model for solid waste bins to make them more automated. Several sensing technologies have been merged and coupled to provide detection of bin states and other required parameters. Although the algorithms and their results seem to be efficient for automatic bin state monitoring, the study lacks remote bin monitoring. In a similar line, Anagnostopoulos et al. [4] have introduced the notion of dynamic scheduling, which is necessary for dustbin cleaning. Top-k query helped cleaning of garbage cans on a first-come, first-served basis. Detecting the kind of waste has yet to be figured out and is only a notion. Furthermore, nothing is being done to address the issue of garbage collection. Zaslavsky et al.'s [5] and Helmy et al.'s [6] analyses on sensor data management would aid in the data separation of different dustbins. The suggested system compensates users by computing points based on weight and trash type inserted using a waste-type recognition system. Suresh et al. [7] explored different approaches to monitor IoT, its capabilities, and basic ideas about creating an application related to information management over the internet. Further, the inclusion of IoT fog nodes into the smart dustbin ecosystem will reduce the delay in cleaning the dustbins [8].

6.3 Proposed Methodology

Rather than utilizing a large number of bins in a disorganized way across the city, one might have a small number of smart bins that are available and inexpensive to release. In this paper, a methodology is proposed for real-time monitoring of the garbage level and its threshold value using several sensors. To control overflow and prevent people from dumping trash outside the dustbin, a smart bin system is suggested, which can detect overflow and inform the appropriate authority. It will also identify wastes being put outside bin and alert people not to dispose of wastes outside with red LED and a buzzer. Additionally, it will also let people know that the dustbin is full and to use next dustbin. Apart from this, a water sensor is there that will detect water in case of rain and will automatically close the door. The data generated by different sensors will be transferred to the control unit and updated regularly via the Wi-Fi module, which will notify the appropriate authorities regarding the status of each dustbin. As a result, the waste collection vehicle will only be dispatched when it is required. Based on which, the optimum route for the garbage collecting van must be identified, reducing fuel consumption, costs, time, and labor. These cost savings and effective waste management may be accomplished by applying this resource optimization. Wet sensor and humidity sensor data will be used to determine if trash is entirely separated or not, which will aid in recycling, disposal, and reuse. To create reports and to do qualitative analysis, data mining will be used. The major objective of the technology that will be deployed is to substitute the city's current time-consuming method, enabling this to become a smart city. The block diagram of the proposed methodology is shown in Figure 6.1.

6.3.1 Hardware and Software Used

The smart dustbin is built on the Arduino UNO board platform. It is connected with a bin ultrasonic sensor (HC-SR04), a Wi-Fi module (ESP8266) 2.4GHz band, register, load cell, humidity sensor, LED, capacitor, Vero board, jump wire, and power supply adapter.

The description of the components used are as follows:

  a. **Ultrasonic sensors:** The ultrasonic sensors generally use the concept that bats use, i.e., sonar. This measures the distance between two objects and offers an amazing range of contact-free distance

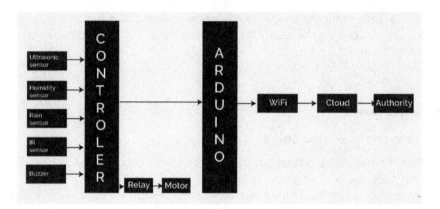

FIGURE 6.1
The block diagram of the proposed methodology.

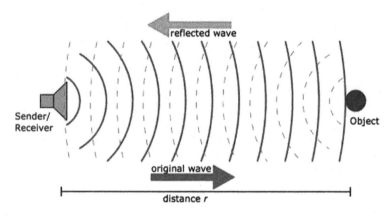

FIGURE 6.2
Ultrasonic transmission and its reflection wave.

ranging from 1 to 13 feet. This process is generally not affected by light or black materials; the soft materials might cause the process to slow down a bit. It becomes complete with ultrasonic transmitting, as shown in Figure 6.2. The ultrasonic sensor is shown in Figure 6.3. The level of garbage in the dustbin will be detected using an ultrasonic sensor. The distance between the sensor and the garbage in the dustbin decides the level of the garbage. VCC (5V), Trig, Echo, and GND are the four pins on this module. Trig is used to transmit a high-level ultrasonic pulse lasting no less than 10s before the Echo Pin detects the returning pulse. The sensor will estimate the distance by calculating the time delay between transmitting the signal and receiving its echo.

It operates by delivering a sound wave (original wave) and receiving a response from the other end (reflected wave), as shown in

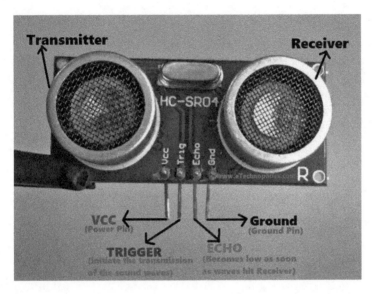

FIGURE 6.3
HC-SR04 ultrasonic sensor.

Figure 6.2. We can calculate the distance(s) by calculating the duration of passage of a sound wave.

d = [Speed * Time];

Time = The time it takes for a sound wave to travel to and from a bin. The speed is equal 330m/s (i.e., velocity of sound)

So, total distance travelled (D) = [d * 2].

Technical specifications

   i. Resolution: 0.3cm

  ii. Effectual Angle: <15°

 iii. Ranging Distance: 2cm–400cm/1″–13ft

  iv. Measuring Angle: 30°

   v. Power Supply: 5V DC

  vi. Quiescent Current: <2mA

 vii. Working Current: 15mA

viii. Frequency: 40Hz

  ix. Range 2cm–4m

b. **Humidity sensor:** A temperature and humidity sensor must be employed to differentiate dry and wet trash (as shown in Figure 6.4). The DHT11 sensor will be utilized for this purpose. Dry and moist waste would be separated based on the output temperature. The

FIGURE 6.4
Image of a DHT11 humidity sensor.

sensor detects ambient air using a capacitive humidity sensor and a thermistor and outputs a digital signal on the data pin.

Every two seconds, the sensor will get new data. With a 2–5% margin of error, it will be accurate for 0–100% humidity measurements and -40 to 80°C temperature readings with a 0.5°C margin of error.

Load Cell: To weigh the dustbin, the load cell must be utilized. It is basically a transducer which produces an electrical signal proportional to the force being measured. It can measure anything from a few micrograms to 200 kg. The output signal is generally in the millivolt range and will require amplification before it can be used. In order to extract quantifiable data from a load cell, the HX711 load cell amplifier must be utilized.

Rain sensors are coupled in order to detect rainfall. A sample of the rain sensor is shown in Figure 6.5. Rain sensors are basically a collection of water sensors linked together by AND gates. They are installed on various sides, and rain is recognized by combining the data from all of the water sensors.

c. **Wi-Fi module (ESP8266):** The ESP8266 is a low-cost Wi-Fi microchip with TCP/IP stacks and the ability to act as a microcontroller. This tiny gadget aids the microcontrollers in connecting to a Wi-Fi network and establishing TCP/IP connections using Hayes-Style instructions. The inexpensive cost and the fact that the module only required a few external components. The ESP8285 is an ESP8266 with 1 MB of built-in flash that enables single-chip Wi-Fi connectivity. The successor of this chip is ESP32 which is shown in Figure 6.6.

FIGURE 6.5
Image of a rain sensor.

FIGURE 6.6
The Esp8266 Wi-Fi module.

d. **Arduino:** Arduino boards are used for establishing the connection and allow management and working of the various sensors of the dustbins. These boards are available commercially in the pre-assembled form. People having Lesser General Public License (LGPL) or General Public License GNU (GPL) are responsible for creating and distributing these boards.

A set of digital and analog pins located within the boards are used for connecting the sensors. The boards also have a USB (Universal Serial Bus) by which users can load programs into the board from

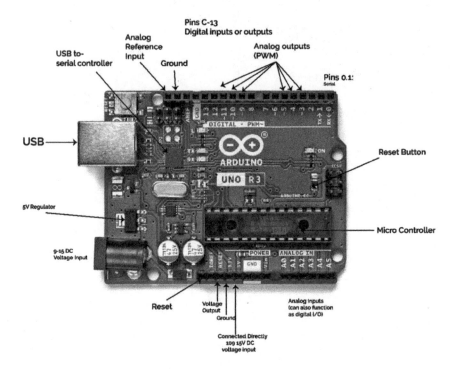

FIGURE 6.7
Arduino Uno Board.

their computers to efficiently manage the working of various sensors. The pin diagram of the Arduino board is shown in Figure 6.7.

i.   Voltage regulator: This stabilizes the processors' DC voltage and controls the input voltage to the Arduino.

ii. The Power USB-Arduino board works well with the help of a single USB connection connected to the USB port of your computer.

iii. Power (barrel jack): By connecting it to a barrel jack, the boards can be easily powered up with the help of AC mains.

iv. Crystal Oscillator: This aids in tackling timing issues. The number on its top is the frequency.

v.   Arduino Reset: There are two options for resetting the Arduino and making it work from scratch, either by using the reset button or by connecting it to the board with an external reset button present labeled as RESET.

vi. Pins: 5V Supply 5 Output Volts.

vii. Pins: 3V Supply 3.3 Output Volts.

viii. GND (ground): These help in grounding the circuit. They are multiple in number.

ix. Win: This pin is used to power up an AC mains power supply from an external power source like the Arduino board.

x. Analog pin: Pins from A0 to A5 can read signals from analog as well as moisture sensors and convert these to a digital value that the microprocessors can process.

xi. Main microcontroller: This is considered to be the brain of the board. ATMEL is the primary manufacturer of microcontrollers. The information available on top of the IC tells you about the board.

xii. ICSP pin: ICSP is an AVR, a small programming header that includes SCK, MISO, MOSI, RESET, VCC, and GND. This is often known as Serial Peripheral Interface, which can be considered an "extension" of the output.

xiii. TX and RX LED-lighting: There are two labels on your board, TX (transit) and RX (receiver). They are given away at various points on the Arduino board. One on virtual pin zero and one indicates all of the serial communication pins, and any other one is TX and RX. While attempting to send serial data, the TX flashes at different speeds. The speed at which it flashes is determined by a variety of factors, including the baud rate used by the board and the RX flashes when it receives data.

xiv. Digital I/O: The Arduino UNO board has 14 virtual pins, six of which are used for PWM, or Pulse Width Modulation output. These pins can be used as enter pins to test logical values, such as zero and one, or as virtual output pins to power modules such as LEDs, relays, and so on.

xv. AREF-AREF stands for Analog Reference: it is used to get an outside reference voltage between 0V to five V because the top restricts entering pins.

e. **PIC16F877A microcontroller:** This is well-known in the industry. It's incredibly easy to use, and it also makes programming easier. One of its primary advantages is it can be rewritable as it uses FLASH memory. As shown in Figure 6.8, it contains 40 pins in total, with 33 for input and 33 for output.

It is versatile, permitting it to be applied in regions wherein microcontrollers have in no way been hired previously, along with microprocessor packages and timer tasks.

- It features a 35-instruction set that is somewhat smaller.
- It can operate at a frequency of up to 20MHz.
- Between 4.2 and 5.5 volts is the operational voltage. If you give it a voltage, it will work.
- It may be irreversibly destroyed if the voltage exceeds 5.5 volts.

FIGURE 6.8
PIC16f877a microcontroller.

FIGURE 6.9
Jump wires.

- It does not have an inner oscillator now.
- The most contemporary every PORT can sink, or supply is around 100mA. Therefore, the contemporary restriction for every GPIO pin of PIC16F877A is 10 milli-amperes.
- It had 4 IC packaging which included 40-pin PDIP, 44-pin PLCC, 44-pin TQFP, 44-pin QFN.

f. **Jump wires:** Jump wires, also known as jumper wires (as shown in Figure. 6.9), are used for solderless bread boarding. They can be manufactured manually, though manually manufacturing the jump wires is a tedious job for large circuits; they can also be obtained in ready-to-use jump wire sets.

No object present – no IR light detected by sensor

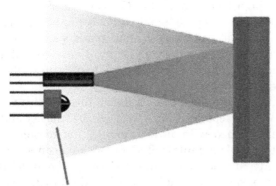

Object present – reflected IR light detected by sensor

FIGURE 6.10
Working of IR sensors.

g. **Infrared (IR) sensor:** An IR sensor is a piece of electrical equipment that detects/monitors infrared radiation. Infrared sensors are linked to detect trash put near the dustbin. When a thing is tossed near the bin, the infrared sensor identifies it, and activates the buzzer.

As shown in Figure 6.10, IR sensors detect light of a specific wavelength in the Infra-Red (IR) spectrum using a specialized light sensor. Using a LED that generates light at the same wavelength as what the sensor is searching for, you can test the intensity of the received light. The light from the LED bounces off an item and enters the light sensor when it is in close proximity to the sensor. This results in a considerable rise in intensity, which may be detected using a threshold.

6.4 Proposed Architecture

The smart IoT-based waste management system design focuses on optimization algorithms and techniques for collecting public garbage. The proposed approach can be adopted in smart cities where people are already overburdened with their demanding schedules and don't have time to waste on administration [9]. If desired, bins can be deployed in a city where a large bin with the capacity to collect solid waste from a single house can be installed. As the first step, the system creates the different sensor nodes as garbage collectors in cities. Each container has a varied storage capacity; therefore, we fill the containers at random. In the second step, before locating the vehicle root, we collect all of the filling ratio measurements from each container and feed them into the genetic algorithm. In the third step, the genetic algorithm will run all of the input populations, and after GA is finished, it will determine the vehicle root depending on the likelihood of container filling. Each container is given four to five attributes as chromosomes, such as container ID, location, capacity, current filling ratio, weight, and so on. We will evaluate the real-time accuracy and compare it to various current approaches after GA has supplied the best path [6].

Dustbins are positioned at various locations inside this structure. In this scenario, garbage bins will be separated into two types i.e., master and slave dustbins. Slave will be equipped with an IoT module, and master with an Arduino Uno along with a Wi-Fi module (ESP8266) for cloud connectivity [9]. Every dustbin is assigned a unique identifier. A database will maintain information about dustbins that should be put in areas based on their associated IDs. The smart, clean dustbins are linked to the internet to provide real-time status updates. Two ultrasonic sensors are positioned at the highest point of the dustbin to minimize incorrect level measurement, while a load cell is put at the foot of the dustbin to feel the weight of the garbage in the dustbin and to decide whether or not the threshold limit is met, a load sensor, as well as an overweight sensor for the rubbish in the bin, and a humidity sensor for wet and dry waste identification are included. Then, for that IR, sensors were installed over the bin's lid to detect waste outside the bin. If there is any trash around, the sensor detects it. When rain is detected, the rain sensors immediately lock the dustbin. At each trash can, all of these sensors are linked to a PIC microcontroller. These sensors transmit data to the controller. Additionally, three LEDs of different colors are used, namely red, orange, and green. When the dustbin is 100% full, the red LED glows, which signifies that the dustbin is 100% full, the green light indicates that the dustbin is empty, and the orange LED signifies the dustbin is 50% filled. Here, a rain sensor is attached to detect rain and close the LED of the dustbin so that rainwater doesn't enter the dustbin. Then the IR sensor detects if some garbage is left outside the dustbin; for that, an installed buzzer buzzes so that people are made of this. Arduino Uno will have to be addressed by every garbage, slave, or master bin with a Wi-Fi module, and Arduino Uno will function as a broker. Arduino Uno's will gather data from sensors mounted to master and slave dustbins, perform noise reduction techniques, and transfer data to the server over Wi-Fi. Arduino Uno must send a message to the server regarding the amount of garbage in a bin, wet and dry waste segregation levels, and the dustbin IDs.

The server matches IDs with the database and will find levels of dustbins at different locations across the city. The algorithm has been developed to check the filling status of dustbins periodically. If the dustbin is filled to its capacity, it will be indicated by LED lights; simultaneously, the encoded signal will be transferred to the dustbin via the broker. Various IoT protocols can be utilized for data transmissions like MQTT or COaP. The collected data in the cloud will be analyzed using different analytic tools to extract useful information. From the collected data, real-time garbage levels can be monitored by the webpage, which will help the appropriate authority to track the exact location and amount of the garbage. The garbage trucks may then discharge the garbage at a specific spot, and the garbage collection vehicle can identify the most efficient path for waste pickup. When the garbage level exceeds a certain threshold, an alarm is sent out for immediate garbage pickup. The data on wet and dry segregation levels will aid in analyzing and refining current waste management methods to improve efficiency. The user will be able to utilize this system effectively thanks to the user-friendly Web GUI and Android applications. The waste collection department will be able to trace the exact position and amount of rubbish by monitoring the webpage. The garbage trucks may then discharge the trash at a specific spot. An innovative solution to the problem of waste management is the Dynamic Routing and Intelligent Transportation System [10].

Figure 6.11 shows the master-slave architecture of the dustbin. The small dustbin acts as a slave and the big dustbin as master. The master dustbin is

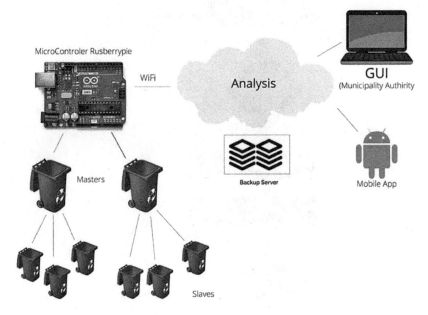

FIGURE 6.11
Overview architecture of smart dustbin.

attached to the Arduino Uno. Arduino is connected to the internet and transfers data of all the dustbins data to the cloud.

Figure 6.12 depicts the project's flow chart. It simply lays out the project's concept. Option starts the flow of the smart waste management system project. Ultrasonic sensors are used to detect the amount of garbage in the bins, and when it reaches a certain level, a message is sent to the appropriate authorities through the Arduino Wi-Fi module so that the dustbin may be cleaned as quickly as possible. The procedure continues indefinitely until the dustbin is cleaned [9].

6.4.1 Architecture of System Components

For better comprehension, this system has been divided into three components that will communicate with one another as shown in Figure 6.13. The

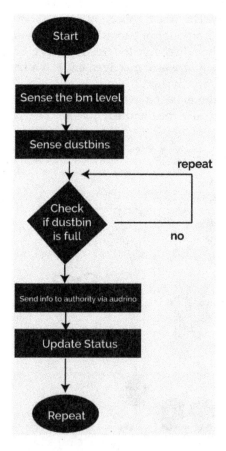

FIGURE 6.12
Flow chart of the system architecture.

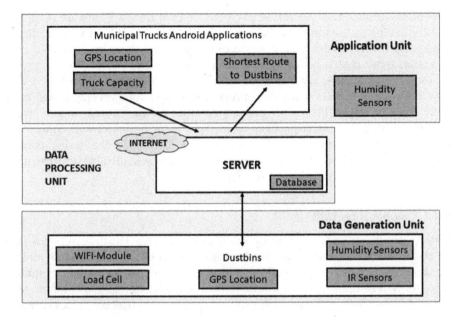

FIGURE 6.13
Architecture of system.

three components are: i. Data Generation Unit, ii. Data Processing Unit, and iii. Application Unit.

6.4.1.1 Data Generation Unit

This unit consists of five sensor-equipped trash cans. These include an ultrasonic sensor (HC-SR04) as a level indication, a humidity sensor, a load sensor to provide an estimate of weight, an IR sensor linked to a buzzer to alert people that they have left garbage outside and near the dustbin, and a rain sensor to prevent the dustbin from opening during rain. The controller (PIC16f877a) is used to examine the values from all of the sensors and send the data to the master Arduino Uno. In order to transfer data, Arduino Uno sends messages and establishes connectivity between the server and the dustbins. An Arduino Uno r3 is a microcontroller board that is used to collect data readings from each dustbin and transfer the data to the server [11].

6.4.1.2 Data Processing Unit

This unit consists of the server that contains an algorithm that determines which bins in the collection list should be updated based on the data received. The algorithm then computes the entire weight from the list, and the variety of vehicles and form of vehicles are selected appropriately. There are several techniques for selecting dustbins for collection, one of which is a cluster-based algorithm in which a collection of neighboring dustbins is treated as a cluster. After that, a list of these clusters that need to be collected is pre-served. These clusters are picked as collecting nodes using a Top-k query-based approach based on the data properties of the vehicles [12]. Based on their capacity, these clusters are chosen as nodes for each truck, and each is assigned a path to collect garbage from the cluster nodes. It will compute the shortest path from the truck's current GPS location to the nodes using Dijkstra's method [13]. This would increase efficiency while also addressing the problem of overloaded trash trucks [14].

6.4.1.3 Application Unit

Two Android applications make up the application unit. The first, depending on the user's GPS position, will offer information about the cluster's clos-est dustbins [15]. Second, the GPS position of the vehicle as well as its total capacity will be obtained by a municipal truck application and, as illustrated in Figure 6.14, the application will provide a route as well as the nodes where waste must be collected for that specific vehicle.

6.4.1.3.1 Product Prototypes

Figures 6.15 and 6.16 show the top and side of the prototype of smart IoT dustbin.

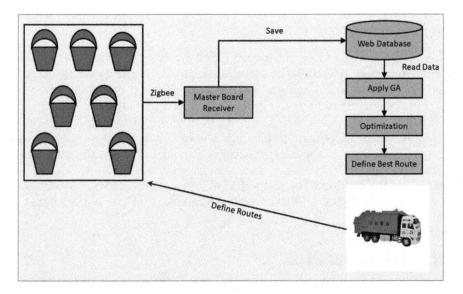

FIGURE 6.14
Overall end to end process of the garbage collection.

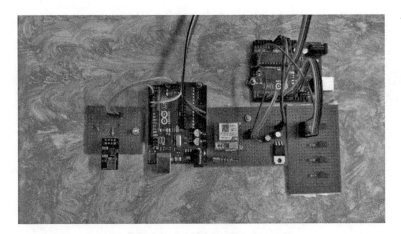

FIGURE 6.15
Top view of IoT Device.

FIGURE 6.16
Side view of IoT device.

6.4.1.3.2 *System UI of Android App*

Figure 6.17 shows the android application UI, where users can log in and give access to current location. Then the application will let the user know about the nearest empty dustbin. The application also helps the authority to display which dustbin needs to be emptied and the shorter path to reach the dustbin.

6.4.1.3.3 *Test Cases and Results*

Software or hardware system testing (as shown in Figure 6.18) involves evaluating the acceptability of a whole integrated system against its defined requirements. It's a part of the black-box testing, which does not really

FIGURE 6.17
UI of android app.

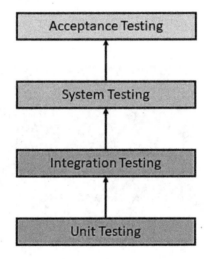

FIGURE 6.18
Different testing processes.

require any prior knowledge about the code or internal structure or implementation [16].

Table 6.1 shows the different test cases and different results obtained. Initially, when the dustbin is empty, a green light glows to indicate people can drop garbage into dustbins; when the dustbin is filled up completely, a red LED illuminates to indicate that the dustbin is full and another dustbin should be used.

6.5 Conclusion and Future Scope

Smart dustbins are indeed a solution for unsanitary environmental scenarios in cities. Using the garbage collection bin, which uses the internet, IR sensor,

TABLE 6.1

Test Cases and Results

Test ID	Test Case Title	Test Condition	System Behavior	Expected Result
01	Empty	Obtaining a full empty length of dustbin	LEDs ON	Same (green LED ON)
02	25% full	Obtaining the 75% empty length	One LED ON	Same (green LED glows)
03	50% full	Obtaining the 50% empty length	One LED ON	Same (orange LED glows)
04	75% full	Obtaining the 25% empty length	One LED ON	Same (orange LED glows)
05	100% full	Obtaining length between 5 to 2 cm as fully filled	One LED ON	Same (red LED glows)

and Arduino, this system is assured as it sends notifications on status. This also alerts the authorities to take proper action on time to clear the litter. This smart bin method is helpful to reduce dependency on manual labour and is a step closer to developing the concept of smart connected cities. It cuts down on time to a large extent, thereby making all the actions more feasible. This project paves the way for a better future that is almost free from land pollution due to proper waste management, and the prevention of many diseases caused this way. The bins can be made more efficient if the segregation of different types of wastes is added to them.

References

1. Parikh, P. A., Vasani, R., & Raval, A. (2017 October). Smart dustbin- "An Intelligent Approach to Fulfill Swatchh Bharat Mission". *International Journal of Engineering Research in Electronics and Communication Engineering (IJERECE)*, 4(10), 6–9.
2. Damakale, N., Rite, P., Wagh, A., & Ansari, S. (2019). IoT based smart dustbin. *International Journal of Scientific Research in Science, Engineering and Technology (IJSRSET)*, 5(6), 513–517.
3. Mamun, Al., Abdulla, Md., Hannan, M. A., Hussain, A., & Basri, H. (2014). Integrated sensing systems and algorithms for solid waste bin state management automation. *IEEE Sensors Journal*, 15(1), 561–567.
4. Anagnostopoulos, T., Zaslavsy, A., Medvedev, A., & Khoruzhnicov, S. (2015, June). Top--k query based dynamic scheduling for IoT-enabled smart city waste collection. In 2015 16th IEEE International Conference on Mobile Data Management (Vol. 2, pp. 50–55). IEEE.
5. Zaslavsky, A., & Georgakopoulos, D. (2015, June). Internet of things: Challenges and state-of-the-art solutions in internet-scale sensor information management

and mobile analytics. In 2015 16th IEEE International Conference on Mobile Data Management (Vol. 2, pp. 3–6). IEEE.

6. Abd Wahab, M. H., Kadir, A. A., Tomari, M. R., & Jabbar, M. H. (2014, October). Smart recycle bin: A conceptual approach of smart waste management with integrated web based system. In 2014 International Conference on IT Convergence and Security (ICITCS) (pp. 1–4). IEEE.

7. Suresh, P., Daniel, J. V., Parthasarathy, V., & Aswathy, R. H. (2014, November). A state of the art review on the Internet of Things (IoT) history, technology and fields of deployment. In 2014 International Conference on Science Engineering and Management Research (ICSEMR) (pp. 1–8). IEEE.

8. Nayak, S., & Banik, D. (2020). Catalyst is important everywhere: The roles of fog computing in an IoT-based e-healthcare system. In *Interoperability in IoT for Smart Systems* (pp. 195–222). CRC Press, Boca Raton, FL.

9. Nayak, S., Dash, A., & Swain, S. Standardization of big data and its policies. *Privacy and Security Issues in Big Data: An Analytical View on Business Intelligence*, 79–107.

10. Medvedev, A., Fedchenkov, P., Zaslavsky, A., Anagnostopoulos, T., & Khoruzhnikov, S. (2015). Waste management as an IoT-enabled service in smart cities. In *Internet of Things, Smart Spaces, and Next Generation Networks and Systems* (pp. 104–115). Springer, Cham.

11. Agrawal, R., & Sharma, A. (2021). Smart bin management system with IoT-enabled technology. In *Soft Computing: Theories and Applications* (pp. 145–153). Springer, Singapore.

12. Anagnostopoulos, T., Zaslavsy, A., Medvedev, A., & Khoruzhnicov, S. (2015, June). Top--k query based dynamic scheduling for IoT-enabled smart city waste collection. In 2015 16th IEEE International Conference on Mobile Data Management (Vol. 2, pp. 50–55). IEEE.

13. Mamun, M. A. A., Hannan, M. A., Hussain, A., & Basri, H. (2014). Real time bin status monitoring for solid waste collection route optimization. In 5th Brunei International Conference on Engineering and Technology (BICET 2014), 11 June 2015.

14. Lincy, F. A., & Sasikala, T. (2021, June). Smart dustbin management using IOT and blynk application. In 2021 5th International Conference on Trends in Electronics and Informatics (ICOEI) (pp. 429–434). IEEE.

15. Kadam, S., Joshi, B., Gada, U., Chaugule, A., & Bhelande, M. (2021). IoT based smart toilet and smart dustbin. *International Engineering Journal for Research & Development*, 6(ICMRD21), 13.

16. Srinivasan, P., Thiyaneswaran, B., Jaya Priya, P., Dharani, B., & Kiruthigaa, V. (2021). Iot based smart dustbin. *Annals of the Romanian Society for Cell Biology*, 25(3), 7834–7840.

7

IoT-Based Smart Waste Management for Smart Cities

Nitika Rani

CONTENTS

7.1 Introduction

Urbanization, better quality of life, and economic development increase the amount of waste generation. According to the World Bank, the world's production of municipal waste is 2.01 billion tons annually. A n average 0.74 kg of waste is generated per person per day (The World Bank). It is estimated that the rate of annual waste generation could rise to 70% from 2016 to 2050, and consequently the amount of waste generation would also rise from 2.01 billion tons in 2016 to 3.04 billion tons in 2050 (World Bank, 2019).

No authentic estimation is available regarding municipal solid waste generation in India. According to the Central Pollution Control Board and Ministry of Urban Development in India, between 2014 and 2015 the amount of waste generation was 52 million tons. Reports given by a task force on energy to the Planning Commission in 2014 gave a rate of waste generation of 62 million tons. On the basis of the total population of India being 440 million in 2017 and the rate of waste generation per capita on daily basis is 450 gm, the municipal waste generation for 2017 comes to 72 million tons, and when per capita generation is reduced to 400 mg then the estimation for municipal waste generation is lower 64 million tons (Ahluwalia and Patel, 2018). It is estimated that out of the total municipal solid waste (MSW), approximately 80% was collected, while only 22% was processed or treated. If waste management methods like separation, collection, conveying, management, and disposal are not done properly, this leads to environment degradation and affects quality of life (Municipal Solid Waste Management Manual, 2016). The rate of municipal waste generation is highest among the biggest six metropolitan areas of India: Delhi, Mumbai, Kolkata, Chennai, Bengaluru, and Hyderabad. Hyderabad produces 4,000 TPD (tons per day) while Delhi produces 9,620 TPD waste, and they constitute 16% of the total urban population of India. MSW generation by these cities constitutes 21% of the total waste generated by all cities in India. Among less populated cities between one and five million inhabitants, Kanpur and Lucknow are the top waste generators from 1,500 TPD to 1,200 TPD (Ahluwalia and Patel, 2018). In terms of per capita waste generators, South Asia, therefore including India, is among the lowest waste generators with 0.5 kg per person per day, while North America (Bermuda, Canada, and the United States) is the highest waste generator with 2.2 kg per person per day. Waste generated by Sub-Saharan Africa is at the lowest rate, with 0.45 kg per person per day (Times of India, 2020). Discarding garbage in public places makes that area unhygienic and also becomes the reason for various types of disease in the locality. This decreases the value of that area due to the bad odors.

A waste management system by means of IoT is one of the latest and most appropriate methods to reduce health-related problems and to improve tidiness in cities (Misra et al., 2018). The combination of information and communication technology (ICT) with advanced progressive projects helps to modify the topography of urban landscapes by making smart cities. Smart cities upgrade the standards of living of the population, add more investment possibilities, and create a healthy and sustainable environment (Kumari et al., 2019). Today, many governments are planning to convert their cities into smart cities. Smart cities mainly include smart economy, smart people, smart living, and smart governance, all of which can be achieved through residents being self-decisive, independent, and smart (Chaudhari and Bhole, 2019). In smart cities, the problem of solid waste generation is very serious and its management should be taken seriously. Two main factors responsible for the

TABLE 7.1

Top Municipal Waste Generator Cities of India, 2016

City	Population (in millions)	Waste Generation (tons per day)
Delhi	19.1	9620
Mumbai	20.0	8600
Kolkata	14.7	6000
Chennai	10.1	5000
Bangalore	10.4	4200
Hyderabad	9.1	4000
Ahemdabad	7.5	2500
Pune	5.8	2300
Surat	5.8	1680
Kanpur	3.0	1500
Lucknow	3.3	1200
Nagpur	2.7	1000
Jaipur	3.5	1000
Ludhiana	1.7	850
Indore	2.5	850
Coimbatore	2.6	850
Agra	2.0	790

Source: State Pollution Control Boards, Municipal Corporation, and UN population estimation.

generation of waste are, first, the growing rates of the population, and, second, standards of living (Ali et al., 2020) (Table 7.1).

Continuous modernization and population increases lead to the generation of large amounts of waste in urban areas. Open dumping is a fire hazard, and causes damage, diseases, economic loss, and environmental pollution. To reduce these problems, large numbers of garbage bins are kept at different sites, and regular checking and collection of waste is done to keep the environment clean *(Saha H. et.al 2018)*. Improper waste management results in overflowing dustbins in smart cities, resulting in various health issues to humans and adverse impacts on the environment (Abba and Light, 2020). Therefore, effective solid waste management is required to minimize these effects. Solid waste management includes reduction, collection, treatment, and disposal of solid waste in a productive way so that human health and the environment are protected. The advancement in technological developments that has led to change in the field of waste management is because of the internet (Pardini et al., 2019). The internet has transformed the world by connecting computers to the World Wide Web, helping communication on a large scale. So in order to manage overloaded dustbins and prevent the causes of fire and other environment related problems, Internet of Things (IoT) is one of the advanced techniques which are very useful in order to deal with the abovementioned problems. Smart municipal waste management includes the uses of IoT, where sensors are placed in dustbins to keep an eye on conditions like fill-level, temperature, humidity, etc. and to also ensure that bins are serviced only when needed; this reduces operational costs and improves customer experience. The structure of IoT consists of multiple devices and appliances which are interconnected.

7.2 Municipal Waste

Waste refers to a substance which is left after its use and is also undesirable, unpleasant, and of no use. It is thrown away after use. It can also be defined as material or objects that are considered to be disposed of or are essential to be disposed of according to the legal norms in the national law. In order to achieve a better life, people move into cities. With the rise in population in cities, the amount of waste generated also rises (Kumari et al., 2019).

There are many sources of solid waste generation, and there is no link between source and municipality. The work of municipalities to manage this waste became the origin of the word "municipal" (Figure 7.1).

FIGURE 7.1
Sources of municipal solid waste.

Waste can be categorized on the basis of its origin (domestic, industrial, etc.), content (organic, glass, plastic, etc.), and its potential hazards (being toxic, flammable, etc.) (Solid Waste, India Water Portal). In municipal waste, municipal means being related to local government, and municipal waste means waste which is taken and treated by the municipality. In India, municipal corporations are in charge of management activities related to public health. Municipal waste includes waste generated by houses, offices, institutes, shops, businesses, gardens, yard wastes, etc. (Agarwal et al., 2015). The whole process of the solid waste management (SWM) system includes main four steps, namely, cleaning, collection, conveyance, and treatment. Out of these, the cleaning and collections are conducted by the municipality corporation, the public health department of the city, while the conveyance and treatment of waste are done by the transportation department of the municipality corporation. The whole city is categorized into different areas, and areas are again categorized into different wards for waste collection and transportation.

Presently, SWM in India is done by taking away waste from residential and industrial areas and transporting it to dumping sites. This is the duty of authorities, usually municipal, to deal with the solid waste generated within their respective areas. Waste collection is purely on a contract basis and is carried out by junk pickers, contractors, and municipalities (Agarwl et al., 2015). The most common methods of SWM are open dumping and landfill as they are cheap, easy, and convenient methods, but they cause adverse impacts to the environment such as environmental contamination and methane gas generation by means of the decomposition of organic matter, which promotes global warming as well other problems like bad odor, waste piling up, etc. (Saleh and Koller, 2019). Mostly human activities lead to waste generation, and the methods used to store, collect, and dispose of waste result in serious risks to the environment as well as to human health (Zhu et al., 2008).

7.2.1 Waste Composition

The types and quality of municipal solid wastes vary from one country to another. Changes can also take place even within the country from place to place due to various factors, such as lifestyle, location, climate etc. The composition of MSW consists of distinct materials generated from different types of activities like intake of food, culture, tradition, way of living, and income. The quantity and composition of MSW vary between different municipalities and the time of the year. Various factors affect the characteristics of MSW like climate, per capita income, and rate of urbanization and industrialization (Singh et al., 2014).

- Paper and degradable organic matter
- Metal, glass, ceramics, plastics, textiles, dirt, and wood etc.

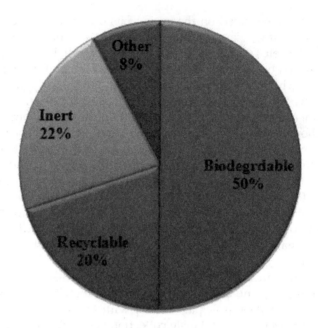

FIGURE 7.2
Composition of municipal solid waste.

- The composition of waste from a particular area also changes, possibly after long intervals or annually due to seasonal variations (Figure 7.2)

7.2.2 Categories of Waste

Municipal waste consists of various types of waste, such as household waste, waste from different types of factories, hazardous waste, demolition waste, agriculture and food processing waste, bio- medical waste, butchery waste, etc. Classification of waste is as follows (Agarwal et al., 2015):

- Biodegradable waste: waste from kitchen: food (cooked and uncooked), green waste like vegetables, flowers, fruits, etc., and paper
- Inert waste: includes C&D waste (construction and demolition)
- Recyclable waste: glass, plastics, cardboard, etc.
- Domestic waste: left-over medicine, e-waste, paint containers, bulbs and tubes, apparatus used for fertilizers and pesticides, batteries.

7.2.3 Waste Management

Municipal authorities play an important role in making society clean, beautiful, and environmentally healthy. Doing their duties efficiently means managing city waste in terms of processing and disposing of municipal solid waste. Methods of waste management may differ from one country to another, between urban and rural areas, and from industrial to residential areas. Waste management includes keeping an eye on all the activities related to waste management. This is from the start, meaning the source of waste generation, then the collection of waste done by junk pickers, after that transportation, and then deposit of waste at its final destination, which can be landfill, incineration, or recycling sites. Thus, if there is any kind of mismanagement or lack of efficient and appropriate solid waste management, the waste generated by citizens' activities results in health risks through disease transmission and the environment being affected (Pardini et al., 2019). Deficient information about the route, location of dustbins, co-ordination between truck drivers, and infrastructure creates negative impacts on SWM, and in order to manage it efficiently, new technologies can be used which help to find the best route to reach the required destinations as well as helping to clean the dustbins in time to prevent overflowing (Özmen and Özsoy, 2020).

In India, municipal waste management is regulated by the Solid Waste Management (SWM) Rules, 2016, which were renamed from the MSW (M&H) Rules, 2000. The MoEF & CC was responsible for revising the 2000 rules.

The salient features of the Solid Waste Management (SWM) Rules, 2016 are:

- Explaining new technologies and options available for municipal solid waste management (MSWM) in India
- Understanding the hypothesis of integrated SWM
- Requirement of state and regional level strategies as well as contributions from cities for MSWM

The amended manual includes all aspects of MSWM, such as drafting of policy, and technological, institutional, economic, and legal aspects.

7.2.4 Concept of Integrated Waste Management

Integrated waste management is based on the three Rs concept: reduce, reuse, and recycle. The aim of an integrated solid waste management (ISWM) system is to minimize the amount of waste being disposed of and maximizing resource recovery and efficiency (Figure 7.3).

According to the integrated solid waste management hierarchy, the methods of waste management, in order of preference, are:

- Reduce waste generation and reuse waste at source: beneficial waste management is to decrease waste generation at different stages, such as during the designing of the product, manufacturing, during wrapping, use, and reuse. Reduction at source helps in the reduction

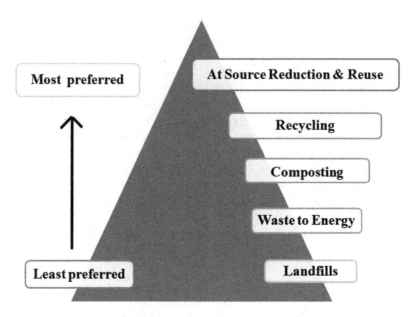

FIGURE 7.3
Integrated solid waste management hierarchy.

of handling, treatment, and disposal costs. It also protects the environment by reducing the generation of leachate and greenhouse gas emissions

- Recycling: the other option is the recovery of recyclable material from waste material and creating a new product. Paper, glass, plastic, and metals are recyclable items. During recycling, disposed material is selected, remanufactured, and converted into useful items

- Waste to composting: composting includes the decomposition of waste organically. Composting of organic waste helps to improve soil health and increase crop production

- Waste to energy: waste to energy includes the production of energy from waste. Incineration is one method of energy generation. This method can be more efficient if proper door-to-door waste collection, segregation and management of inert materials, and correct pre-treatment of two separate energy-generating parts are followed (Surapaneni and Symala, 2018)

- Waste disposal: this is the least preferred waste disposal method and includes the disposal of waste, mainly comprising of inert material, and is done by means of sanitary and lined landfills which can be constructed according to the SWM Rules, 2016 (Municipal Solid Waste Management Manual, 2016)

7.3 Waste Management by Using IoT (Internet of Things)

Using IoT with smart sensors in waste management increases operational efficiency, reduces costs, and enhances customer satisfaction (Sharma, 201). IoT technology is one of the most important applications and well-planned mechanisms to construct smart cities. Due to the continuous increase in populations, smart cities are facing the major problem or challenge of the continuous increase in waste generation (Khoa et al., 2020). Population explosion and an upgrade in the way of living are two main reasons for the generation of solid waste in large amounts, both in rural and non-rural areas of the country (Agarwal et al., 2015). In order to manage the waste, dustbins are provided by the government, people dump their waste in these dustbins, and they are taken away by the municipal corporation (MC). Due to the large population numbers, we find dustbins overflowing at various places in cities, causing unhygienic environments and bad odors and resulting in human illness. It is difficult to stop waste generation, but a smart system can help to get a real-time status of the dustbins. This chapter helps to understand the new and innovative methods that can be used to make cities healthy and clean. These methods or systems are based on the regular monitoring of the levels of waste in garbage bins and connect the dustbins and the waste collector truck drivers by means of organized web-based structures that keep an eye on the status of the dustbins (Abba and Light, 2020) (Figure 7.4).

With the advancement of the internet, smart sensors, technologies, software, interconnection of machines, and communication between objects without the involvement of humans, IoT is the new and latest exemplar for municipal solid waste management. IoT can be defined as an effective and

FIGURE 7.4
Overloaded dustbins.

global network infrastructure, in which smart things, subsystems, and other physical and virtual entities are identifiable, self-governing, and self-configurable (Pardini et al., 2020). IoT includes various physical and electrical instruments, updated vehicles, buildings, and tools, and all these devices and instruments must be linked with each other by means of the internet to exchange data between each other (Ali et al., 2020).

Two main problems related to waste collection are:

- Scheduling: when to collect waste from dustbins
- Routing: selection of route by waste collectors

IoT-based municipal waste management includes management of waste by using sensors in refuse bins to check the fill-level, temperature, and tilt to improve waste collection, and servicing the bin is provided when needed which helps in reducing pollution, fuel consumption, etc. It also helps collectors to manage their daily and weekly schedule to pick up the garbage (Chaudhari S. and Bhole V., 2019).

7.3.1 IoT-Based Municipal Waste Management Methods

The aim of IoT-based smart systems is to take care of waste management in smart cities. In some systems, a signal is sent to the worker by means of sensors through a server, then the worker travels to the required destination to collect the waste. In other methods, systems consist of sensors, micro-controllers and Wi-Fi networks which are used in garbage management systems in order to clean the bins in time when they reach their maximum waste levels. If garbage is not removed within good time, information is sent to higher authorities, and appropriate action is taken (Kadus et al., 2020. One of the similar proposed systems includes the measuring of garbage levels inside the bin, and when the bin is full it sends a signal to the municipality, based on the type of garbage, i.e., household waste, paper, glass, or plastic. This system consists of a sensor and microcontroller, and everything is connected to a device which stores data for future use (Mustafa M.R. and Azir K.N.F, 2017).

7.3.2 Steps Involved in IoT-Based Municipal Solid Waste Disposal Methods

- Relevant hardware that indicates the fill level of waste in the bins: garbage bins with ultrasonic sensors are placed at each location in which sensors indicate the extent of waste in the bins at different levels
- Module on central cloud server: up-to-date information regarding the status of garbage bins is received by this module and is

continuously displayed on the web application. This also sends messages to the client side (municipal corporation, garbage collector truck drivers, etc.) through a mobile application

- Use of app on phone to indicate the location and present waste level of bins on geographical area map: by means of mobile applications, truck drivers hired for garbage collection get accurate information regarding the level of waste in garbage bins and if the garbage bins are full or about to be full.
- Creation and layout of shortest route for garbage collector truck driver to reach the fully filled garbage bins: to empty the fully filled bins within time, the garbage collection truck needs a short route to reach the relevant bins, and so the shortest path for truck is calculated. Management of route development is done on a central cloud server (Chaudhri S. and Bhole V., 2019)

7.3.3 Working of Any IoT-Based Smart Bin Based on Three Modules

IoT-operated perceptive waste collection and the appropriate system is accomplished with the help of different equipment, application software/programming tools, and other IoT devices. Numerous technological modules and sensors are integrated in this system. The function of the system is based on three modules.

7.3.3.1 Sensing Module

The sensing module consists of different components that coordinate with each other as a system. The sensing module includes the Arduino Uno microcontroller, sensors, Wi-Fi network, and solar power operated battery (Figure 7.5).

a) Sensors: in order to check the level of waste in bins, ultrasonic sensors are used. Based on the space inside the dustbin, there are three levels:

FIGURE 7.5
HC-SR04 ultrasonic sensor.

i)   Empty level: this is the initial level of waste in the dustbin, and the system shows this level when dustbins are emptied and unfilled by the garbage truck driver within time.

ii) Half level: this is the next level of waste in the dustbin, and it helps to identify how much time a dustbin will take to fill up. When truck drivers go to take waste from full bins, they can also pick up the waste from these half-filled bins. Emptying half-full bins helps to save fuel and the costs of running the waste truck, and it also helps in the control of air pollution.

iii) Full level: when the sensor triggers "full", status of this level is received by the system and notification is sent by the system to the municipal authority and truck driver.

Different sensors can be used for different parameters in dustbins, for example the temperature sensor and the humidity sensor, the latter of which is to determine the extent of humidity in the debris in order to prevent unfavorable events happening, like fire. The weight sensor is to determine the weight of waste collected from the bins so that predictions for future waste generation from a particular area can be made (Figure 7.6).

b) Arduino Uno microcontroller: by means of a microcontroller, data received from the sensors is sent to a server by means of an internet network. Many types of microcontroller are available and deciding which microcontroller to use depends upon its storage capacity, processing speed, use of electricity, and price. Use of the Arduino Uno-type microcontroller is one good option. Arduino is a computer which is intentionally designed for people who are not experts in electronics, engineering, or programming. It is inexpensive, cross-platform (it can run on Windows, Mac OS X, and Linux), and easy

FIGURE 7.6
Arduino Uno microcontroller.

to program (Chaware et al., 2017, Srikanth, C. et al, 2019). Arduino manages the overall communication. The Arduino Uno controller receives information from the different sensors, and transmission of this information to the server is done via internet networks and services (Ali et al., 2020)

c) Data collected through various sensors are sent to the server through wireless networks. Wi-Fi is used as a network interface

d) Battery: wireless technology, data received by sensors, and forwarding it on are directly influenced by energy consumption. A solar-based charging battery is used to fulfill the energy requirement

7.3.3.2 Storage

The storage module includes the storage of data received from sensors in the form of a database so it can be retrieved easily. This storage can be used to estimate the waste level and generation in the bin monthly or annually.

7.3.3.3 User Module

a) Waste data collection: after getting information about the level of bins, and if the bin is full, the waste truck responsible for that site receives a message and collects the waste

b) Display: the latest information about the bins is displayed on an screen and shared with other devices such as smartphone, tablet, etc. with data access, helping the waste truck to get the updated status (Nirde K. et.al, 2017) (Figure 7.7)

7.3.4 Other Methods to Control the Overflowing of Dustbins

Other methods include Data Gathering Layer (DGL), Data Processing Layer (DPL), and Data Demonstration Layer (DDL).

In the Data Gathering Layer (DGL), readings are taken from waste bins with the help of different types of sensors. The readings from the sensors are sent to the Data Processing Layer (DPL) through a microcontroller (Raspberry Pi module) for processing. Assessment of temperature level, humidity level, and volume of waste inside the bin is done by the Smart Bin Model. The MQTT (Message Queuing Telemetry Transport) protocols, which transport messages between devices, take care of messages and communication between the microcontroller sub-system and the control desk. The Data Demonstration Layer (DDL) consists of the Alert Notification Sub-System and control desk. The Alert Notification Sub System (ANS) includes various types of signals for temperature level, humidity level, remaining waste level inside the bin, and a buzzer or alarm to make the waste collectors aware of the status of the waste bin (Figure 7.8).

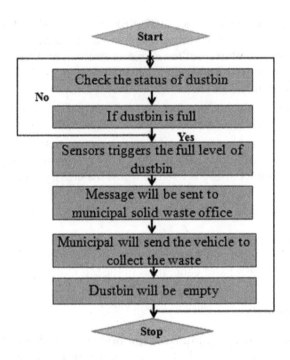

FIGURE 7.7
Diagram showing municipal waste management by IoT.

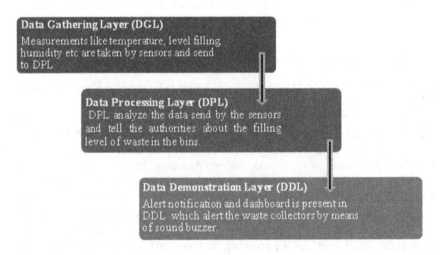

FIGURE 7.8
Other methods for waste management by IoT in brief.

In this method, different kinds of sensors, such as DHT-11 and HY-SR05 are used to accumulate the information about the bin, which is equipped with the Raspberry Pi module. In order to measure the humidity and temperature of the residual waste in the waste bins, the DHT-11 sensor is used. The filled level of the waste bins is estimated by an ultrasonic distance sensor (HY-SR05). Then the information sent by the sensors is analyzed by the microcontroller, Raspberry Pi with Node-Red, which helps to make future projections. If the findings are at a critical level, the information will be sent to the authorities about the waste that needs to be collected and from what place (Kumari et al., 2018).

In addition to the above explained methods of waste management by means of IoT, some other methods are also available as follows:

RFID (radio frequency identification) technology: this technology is mainly for food waste collection. In RFID, RF means "radio frequency" and ID means "identifier". Radio Frequency Identification (RFID) refers to contactless technologies which detect people or objects through radio waves (Kumar N. et al., 2016). In order to locate the bin, scanning from RFID requires the lid of the dustbin to be closed.

Automation of bins: in order to prevent the overflowing of bins, a conveyor belt helps to close the lid of bins when they get completely full. This can be done by means of RF technology (Chellam et al., 2020)

Currently, there are several companies such as Bigbelly, Sensoneo, and Ecube Labs aiming to create smart waste management systems so that cities can develop processes and contribute towards making society healthy and the environment clean. Bigbelly provides waste and recycling systems by using smart systems with cloud-based platforms to manage the solid waste. Sensoneo uses smart sensors and smart analytics for smart waste management by monitoring waste management processes, and optimizing waste collection routes, frequencies, and vehicle load. In addition to this, Sensoneo also provides an application which shows all the smart bins of a city in a city map, which helps to find the closest smart bin as well as the shortest path to reach that bin. Lastly, Ecube Labs offers eco-friendly waste management solutions for smart cities (Özmen and Özsoy, 2020.

7.4 Benefits of Smart Waste Management

Out of the various environmental problems faced by cities, the problem of waste management is of utmost importance because of its adverse effects on the health of people living there (Shyam G. et al., 2017). This chapter helps to understand the different methods of managing overflowing bins in smart cities in a smart way. Overflowing garbage containers impact health and air pollution. Picking up and handling overflowing garbage by waste collection staff is also of great concern, as this can transmit various infections or chronic diseases to them. The latest methods provide up-to-date information regarding the status of dustbins so that the waste collectors can pick up the waste on time (Smart Waste Collection Report by Navigation Research) and also help the waste collector to find the shortest path to reach the particular bin. Proper solid waste management processes (generation, collection, transportation, and disposal) enable the minimization of greenhouse gas emissions, thereby preventing environmental pollution and enhancing the health of citizens.

7.5 Future Prospects

In the future, according to various studies, appropriate sensors can be used in the interior of waste bins to measure their toxic gas levels, radiation levels, and to measure the weight of collected solid waste. Research should be done to detect the suitability of integrating waste
management systems in small towns and their nearby villages for efficient MSWM. Future prospects also include the low cost of the system including development as well as maintenance. Furthermore, the capability of the IoT-based system can be improved by using accessible GIS data without making any speculations.

7.6 Conclusion

Due to increases in the amount of waste generated, we require a more strategic, efficient, and better framework in order to manage waste-related problems. To prevent any kind of epidemic situation as well making smart cities healthy, economically and environmentally, we need a very well-planned technology-based solid waste management system and also strong implementation of this in our country. This chapter explains the different types of smart ways to manage solid waste in smart cities. These methods use IoT (Internet of Things). IoT-based efficient waste management systems take care of all garbage bins located everywhere in the city in a smart way, with characteristics like resource management: inexpensive and time saving. A smart bin increases recycling rates with the help of IoT systems and their collected data. As a result, smart bins are seen as an eco-friendly solution for smart cities.

References

Abba, S. and Light, C.I. (2020). IoT-based framework for smart waste monitoring and control system: A case study for smart cities. In Presented at the 7th Electronic Conference on Sensors and Applications, 15–30 November 2020. Available online: https://ecsa-7.sciforum.ne.

Agarwal, R. et al. (2015). Waste management initiatives in India for human well being. European Scientific Journal, 2015, 105–127.

Ahluwalia, I. and Patel, U. (2018). Solid Waste Management in India: An Assessment of Resource Recovery and Environmental Impact. *Indian Council for Research on International Economic Relations.*

Ali, T. et al. (2020). IoTBased Smart Waste Bin Monitoring and Municipal Solid Waste. *Arabian Journal for Science and Engineering,* 45, 10185–10198.

Chaudhari, S. and Bhole, V. (2019). Solid waste collection as a service using IoT solution for smart cities (conference paper). In 2018 International Conference on Smart City and Emerging Technology (ICSCET). https://www.researchgate.net /publication/329061124.

Chaware, S.M. et al. (2017). Smart garbage monitoring system using Internet of Things (IOT). *International Journal of Innovative Research in Electrical, Electronics, Instrumentation and Control Engineering,* 5, 74–77.

Chellam, S. et al. (2020). Municipal waste handling using IoT. International Research Journal of Engineering and Technology (IRJET), 7, 4027–4034.

Composition and Quantity of Solid Waste. (n.d.). http://cpheeo.gov.in/upload/ uploadfiles/files/chap3.pdf.

Kadus T. et al. (2020). Smart waste management system using IOT. International Journal of Engineering Research & Technology (IJERT), 9, 738–741.

Kumar N. et al. (2016). IOT based smart garbage alert system using Arduino UNO. Conference paper.

Kumari, P.K.S. et al. (2018). IOT based smart waste bin model to optimize the waste management process, 48–57. Retrieved from https://www.researchgate.net/ publication/330511598.

Management system for smart cities. (n.d.). Arabian Journal for Science and Engineering, 45, 10185–10198.

Municipal Solid Waste Management Manual. (2016). Ministry of Urban Development.

Nirde K. et al. (2017). IoT based solid waste management system for smart City. In International Conference on Intelligent Computing and Control Systems, Madurai, India.

Ozmen, G. and Ozsoy, A. (2020). Waste management optimization by using IoT (project research report). https://www.researchgate.net/publication/345941365.

Pardini, K. et al. (2019). IoT-based solid waste management solutions: A survey. Journal of sensor and Actuator Networks, 8.

Pardini, K. et al. (2020). A Smart Waste Management Solution Geared towards Citizens, MDPI.

Saha H. et al. (2018). Waste management using Internet of Things (IoT), Conference Paper. https://www.researchgate.net/publication/320596482

Saleh H. and Koller M. (n.d.). Introductory Chapter: Municipal Solid Waste, Intech Open.

Sambyal, S. (2016). Government notifies new solid waste management rules. Down To Earth.

Shyam G. et al. (2017). Smart waste management using Internet-of-Things (IoT). In Second International Conference On Computing and Communications Technologies (ICCCT'17).

Srikanth, C. et al. (2019). Smart waste management using Internet-of-Things (IoT). International Journal of Innovative Technology and Exploring Engineering (IJITEE), 8, 2518–2522.

Surapaneni, P. et al. (2018). Solid waste management in smart cities using IoT. International Journal of Pure and Applied Mathematics, 118, 635–640.

Zhu, D. et. al. (2008). *Improving solid waste management in India. A Sourcebook for Policy Makers and Practitioners*, World Bank Institute, WBI Development Studies. The World Bank.

8

Serverless IoT Architecture for Smart Waste Management Systems

Rahul Singh, Bholanath Roy, and Vipin Singh

CONTENTS

8.1 Introduction

Compared to past decades, the present way of living is changing continuously, with rapid population growth, urbanization, and industrialization being significantly evident. Substantial economic growth has greatly contributed towards changing patterns of living, which in turn leads to production of municipal solid waste (MSW). This generated waste ultimately has an undesirable impact on the environment as well as on public health. The composition of MSW generally contains (a) organic waste e.g., food waste, inert material, paper, cardboard, textile, and wood, etc. and (b) inorganic waste e.g., construction waste, glass, leather, metals, plastics, Thermocol, etc. (Palanivel and Sulaiman, 2014). It further contains some toxic and hazardous substances such as paints, pharmaceuticals/medicines, syringes, pesticides, sanitary napkins, etc. Approximately 2.1 billion tons of MSW were generated in the world in 2016, which would lead to ~3.4 billion tons by 2050 and where about 0.24 billion tons of plastic waste would be produced globally (Sharma and Jain, 2020). It could be observed that annual MSW generation would increase by 70% by 2050. Presently about 6% of waste is composted and ~14% recycled, whereas ~35–40% of waste is not managed appropriately and found randomly dumped or disposed of via open burning (World Bank, 2018). Appropriate management of MSW is essential for sustainable development and to build a livable environment, but it is a big challenge for several developing countries as well as cities. Along with the changes in living patterns, this is an era of innovations. There are several technological developments in the field of engineering throughout the world that address the issues of various sectors. One of these developments is the Internet of Things (IoT), which helps to automate any particular field at present. As waste management is a regular activity that needs a lot of manpower on a daily basis, it naturally has impact on efficiency of work, accuracy, budget, social aspects, and human health. Alternatively, IoT-based smart waste management systems could reduce the costs of waste management and provide real time information to increase efficiency and develop further strategies. There is a requirement for physical infrastructure in this automated system, and with that comes the associated high costs for the establishment and maintenance of servers. In such cases, serverless IoT plays an important role and reduces the costs of establishment and maintenance.

8.2 Waste Generation and Management

The rate of generation of MSW varies widely among countries in the world and depends mainly on level of income, climate factors, social and cultural pattern, etc. It has been reported that high income countries generate approximately six times more MSW as compared to low-income countries. Per capita MSW generation for Brazil was found to be ~300–600 kg/year in 2012 (UNEP, 2015). India is developing and has the fastest growing economy (sixth largest in terms of GDP and third largest in purchasing power parity) (Gramer, 2016). India is the second most populated country, where ~377 million live in urban areas and produce ~62 million metric tons (MT) of MSW per year. This generation of MSW is increasing 5% yearly and is expected to be 165 MT per year by 2031. Waste generation per capita was found to be ~350–485 grams/day (CPHEEO, 2018). Composition of solid waste generated in India is shown in Figure 8.1, and city waste generation is provided in Figure 8.2 for the years 2000–2016. Proper handling of such huge quantities of waste at a nationwide level is critical, as a lot of effort is required for collection, transportation, treatment, and disposal in order to maintain environmental and public health hygiene. Presently ~22–28% of MSW is being processed and treated, and untreated waste may require ~1240 hectares of land per year for its disposal through landfill, further increasing to 66,000 hectares/year by 2031 (Paulraj et al., 2019).

Shekdar (2009) reported that per capita waste generation was found to be highest for Hongkong (~2.2 kg/day/person) and lowest for Nepal (0.2

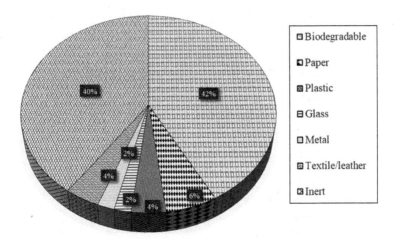

FIGURE 8.1
Composition of Indian solid waste (Shekdar, 2009).

to 0.5 kg/day/person). The integrated approach for MSW management in Asian countries consists of collection systems for MSW and transportation, landfilling, and processing systems for volume reduction or energy/material recovery. There are several methods for processing MSW, such as incineration, pyrolysis, anaerobic digestion, gasification, etc.

The management of MSW is of serious concern in India as well as worldwide. Several developed countries are focusing on readymade solutions with the use of the latest technologies and automation in this sector. IoT is one solution that can reduce human resources for efficient management of MSW.

8.3 Internet of Things (IoT) Era and Its Application in Waste Management

John Romkey, in 1990, was the first person who tested TCP/IP protocol (Khvoynitskaya, 2019). After one year, Cambridge University scientists created a model for a web camera to monitor a staff coffee pot at their computer lab. They programmed the web camera to take a picture of the coffee pot three times per minute and send that picture to the local computer so that any one of them would know whether coffee was available or not. In 1999, Kevin Aston coined the term the "Internet of Things" in his presentation for Procter. He described IoT as a technology that connected many devices via radio frequency identification (RFID) for supply chain management. IoT has many advantages (Brous et al., 2020) that can improve services as well as analysis of historical data. If service improvement is to be focused on,

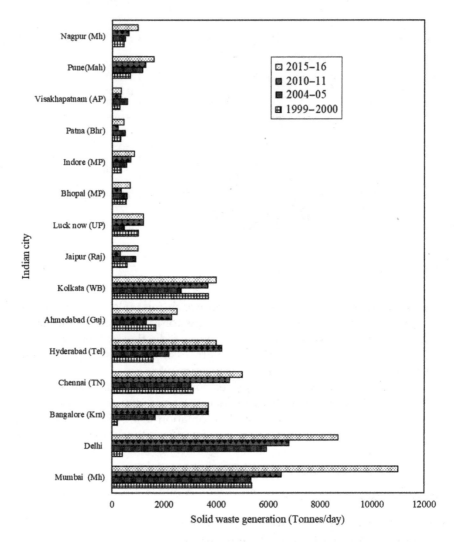

FIGURE 8.2
Solid waste generation in Indian cities between 2000 and 2016 (CPCB, 2018).

automation based on that data with the help of IoT real-time sensor data is efficient and enhances productivity. In the case of historical data, there are different types of sensors by which different types of data collection are possible in a single place. In addition, this type of collected data is used for historical analysis and machine learning algorithms. Some features of IoT are as follows:

(1) **Easy access:** helps to easily access location and mobility and makes certain tasks more manageable.

(2) **Heterogeneity:** IoT integrates different technologies such as tracking and identification. It also uses wired and wireless sensors, Radio Frequency Identification (RFID) technology, latest protocol, and ZigBee technology (Chen and Jin, 2012). IoT creates heterogeneity, which means the different types of sensors and technology work together to collect more valuable data than traditional methods.

(3) **Better time management:** sensors are fast in nature, and we do not use an intermediary between sensors and the cloud server. That's why it will take a little time to reflect sensor data on the server. That is the power of serverless IoT by which we can visualize data and make decisions in real-time. It doesn't take extra time to collect and process the data. So, time management is better. It also cuts associated costs.

(4) **Mobility:** as IoT sensors are small in size and need a low power supply, a small battery can be used to operate them, creating mobility in our system. Many projects and systems where a fixed power supply is not possible can avoid these difficulties with IoT. IoT needed internet, and nowadays internet can also be used by mobile devices to make IoT more usable and reliable (Esquer et al., 2017).

The IoT-based waste management system is designed using various components (Keramidas, 2016; BEHRTECH, 2020) such as the following:

(a) **Temperature sensors:** to sense temperature. These sensors measure the amount of heat energy from where sensors have been installed and later converts that heat energy into data

(b) **Humidity sensors:** to measure the humidity of the atmosphere. They measure the amount of water vapor present in the air

(c) **Pressure sensors:** these are used to sense pressure. They take pressure as an input and convert it into an electrical signal., and then that signal into data

(d) **Proximity sensors:** to detect nearby objects without contact. Such types of sensors often emit electromagnetic fields, and if any entity or person comes in that range, it senses and gives a signal in the form of data

(e) **Gas sensors:** to detect changes in air quality. In other words, we can say that it detects the presence of various gases

(f) **Accelerometers:** to detect the acceleration force of an object, i.e., the rate of change velocity of an object within a certain time

(g) **Optical sensors:** used to convert light into an electrical signal

(h) **Smoke sensors:** detect smoke particles (airborne particles) with some gases in the air. They can be used for fire safety in any location

(i) **Actuators:** used to take certain actions based on specific input signals. They are different from sensors

(j) **Arduino:** an open-source microcontroller-based electronic circuit board having multiple variants with limited memory. It is used as a prototyping platform to connect sensors and actuators. It can be connected to a computer via a USB port and can be programmed

(k) **Raspberry Pi:** a small-sized computer board having a complete operating system. It is used to run any type of application because it has the same features as Arduino in addition to the operating system. The operating system makes it easy to install dependencies (Sethi and Sarangi, 2017)

Misra et al. (2018) used an ultrasonic sensor in a waste bin to find the level of waste in the waste bin. They discovered that the ultrasonic sensor has long-distance ability to sense; that is why ultrasonic was used instead of an IR sensor. The IR sensor is also affected by sunlight. In addition, they find the intensity of biogas levels in municipal areas. Sensor data is transferred over the internet to the server, where it will be processed. After that, monitoring will occur. And according to the level of waste in the waste bin, the relevant waste bin that needs to be picked up will be selected. This research was based on specific scenarios rather than based on daily collection data. Daily collection data and historical data were used to forecast future states and availability of vehicles near the site. An ammonia and hydrogen sulfide gas sensor was also used in the waste bin. The MT-135 sensor was used to assess air quality and dangerous gases. Wi-Fi serial module ESP 8266 was used to transfer data to the server over the internet.

One of the good solutions to smart waste management is Bigbelly (SB, 2021). Initially, it was a solar-powered garbage compactor manufactured by US company Bigbelly Solar. The goal was to reduce the use of fossil fuels. Originally, it was used for public spaces, universities, and parks. Bigbelly uses the Internet of Things. It has been deployed in over 50 countries and is recognized by the C40 Cities Climate Leadership Group. It offers to improve quality of life and increase productivity. It has good features like GPS, is self-powered with solar technology, therefore it is possible to connect with the internet at any time. A similar study was reported for another commercial waste management solution (CH, 2021). The researchers used secure web and IoT sensors to report and visualize real-time data. Based on the location of containers that require servicing, the best route was found. The sensors also indicated whether the container was full or not and generated a smart route to the driver. Bine (2020) also reported another commercial waste management system. This researcher used big data and IoT to identify the waste and after identification, the system will sort the waste according to type. This system analyzes the waste before collection so that after sorting, less space will be taken up than the original size of the load. A centralized system was used to control all the sensors and monitor processing in real time using wireless communication to optimize the logistics.

8.4 Role of Servers in IoT-Based MSW Management Systems

In order to make IoT more valuable, a server is needed. This is typically a hardware infrastructure that works as the heart of an entire network (Figure 8.3) for hosting. These servers are responsible for collection of generated data, connected to IoT devices, and sensors of smart waste management systems. These are either localized or centralized, with a large capacity for memory and processing power to operate an IoT network causing an immediate response. The server saves data and processes it to make usable information for some decisions-based output in the form of signals or graphs or sends data to mobile applications. Many types of servers exist at the moment that can contribute greatly to our needs. In other words, when an IoT application in a waste management system is not connected with a server, it is impossible to work on historical data, leading to delays or irregular decision making. There are also many kinds of real-time data that we can observe by using the mobile app, but that data needs to be saved on the server first (Aman et al., 2018).

8.4.1 Physical or Dedicated Server IoT Systems

This is a traditional or conventional type of server that is basically dedicated computers designated for private use. This server is usually built by individuals using hardware that is arranged in a specific configuration to fulfill a certain requirement. It has physical components like random-access

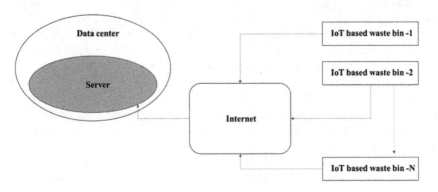

FIGURE 8.3
IoT network model for smart waste management (Aman et al., 2018).

memory (RAM), a central processing unit (CPU), a graphics processing unit (GPU), motherboard, hard drive, and networking card etc. It is powerful and customizable, but certain costs associated with hardware, space, and maintenance would be involved. The owner has access to the complete system in dedicated servers; thus, it provides a high level of security. These servers are not very user friendly, therefore require a skilled person to operate and manage them. Another consideration is that the setup cost of physical servers is relatively high, and if any hardware fails for any reason, the server goes down, which is a big drawback during operations of certain tasks. Sometimes the server is required for a certain task, thus after performing that task, the associated hardware becomes useless. Hence, based on some of these drawbacks of dedicated servers, cloud-based servers were developed as an alternative option, which do not require physical establishment and are also maintenance free for users (Abidi and Singh, 2013; Patra and Barik, 2015).

8.4.2 Serverless IoT Systems

According to Baldini et al. (2017), serverless systems are a new and emerging field which makes applications and services smoother and faster. They were popularized by Amazon in 2014 with Amazon Lambda, after that Google Cloud Functions in 2016, later Microsoft Azure Functions, and then IBM OpenWhisk was developed. Google reported that over a five-year period, i.e. 2016 to 2021, the popularity of searches using the keyword "serverless" increased. One of the most important concerns that people have is about the meaning of serverless. It doesn't actually mean "without a server"; this is a misconception, as it is a cloud-based server. For these servers, there is no need to set up a server of our own; it is purchased from companies or third-party service providers who provide facilities to access their server space with the help of the internet and where the actual location of the server is not known. Also, developers have no concern about server maintenance, scalability, monitoring, and fault tolerance. These responsibilities are taken care of by service providers. In serverless architecture, on the bottom layer, there are many sensors deployed at the endpoint. This means these sensors directly interact with the object or environment where automation is required. Such sensors are used to sense and generate data. These types of sensors are attached to a controller (local computer) that has the ability to take input from sensors in the form of data. There are many controllers in the market, such as Arduino and Raspberry Pi, which are most widely used for educational purposes (Abidi and Singh, 2013).

Arduino and Raspberry Pi have their own computation power and memory (Figure 8.4). Both can collect sensor input in the form of data stored, processed, and manipulated according to need. There are different ports for different types of sensors to take input. Output ports are also available. Most

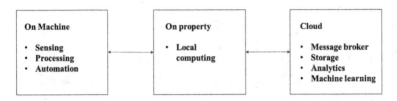

FIGURE 8.4
Typical three-layer IoT architecture (Todd D., 2019).

of the hardware and connections are only required to connect such sensors with controllers because after this layer, hardware, connection wires, and maintenance will be zero for the end-user. These controllers are directly connected to the internet with a wired or wireless connection. A Wi-Fi module could be used for Arduino to connect it to the internet or inbuilt Wi-Fi in Raspberry Pi. Therefore, it doesn't require an external hardware module to connect with the internet. Also, via the internet, our entire network is directly connected with a cloud service provider that manages the rest of the responsibilities, like data collection in proper formatting, data storage, data representation, or generating triggers for other actions (Truong and Dustdar, 2015; Wang et al., 2019).

In Table 8.1 the comparison of serverless platforms of five different companies – Amazon, Microsoft, Google, Oracle, and IBM – has been shown. Their corresponding serverless platforms are AWS Lambda, Azure Functions, GCP Cloud Functions, Oracle Cloud Functions, and IBM Cloud Functions respectively. The Node.js is used by major serverless providers, Microsoft Azure only uses typescript, and GCP Cloud Functions only uses Visual Basic (Sethi and Sarangi, 2017). Each of the above supports custom runtimes. These have concurrency limits, which is essential in terms of availability because they handle a number of server requests that are generally hidden. When it comes to resource memory, this is consumed by code and resources at the time of the active state, each of them providing a minimum of 128 MB and a maximum of at least 1 GB. As an exception, AWS Lambda provides a maximum of up to 10 GB. When it comes to cost, there are not many differences among them, as all are near about the same and initially provide some free usage. The most commonly used runtime languages for serverless applications based on a survey by IBM with 391 users with multiple selection options (IBM, 2021b) are JavaScript, Java, Python, C++, C#, Ruby, .NET, PHP, Node.js, PowerShell, Go, TypeScript, Swift, .NET Core, Bash, F#, Rust, etc. There are various other applications of serverless systems, such as customer relationship management (CRM), data analytics/business intelligence, finance database applications, HR applications, engineering streaming media applications, enterprise resource planning, etc. (IBM, 2021b).

TABLE 8.1

Serverless Services Comparisons (Tropeano, 2019; Harvey 2020; Rifai 2021; Butusov, 2021; Heller, 2021; IBM Cloud, 2021; IBM, 2021a; EOCF, 2021; OFF, 2021)

Services and Product Name	Amazon	Microsoft	Google	Oracle	IBM
Product for Serverless	AWS Lambda	Azure Functions	GCP Cloud Functions	Oracle Cloud Functions	IBM Cloud Functions
Supported Languages	C#, Go, Java, Node.js, PowerShell, Python, Ruby	C#, F#, Java, Node.js, PowerShell, Python, TypeScript	C#, F#, Go, Java, Node.js, Python, Ruby, Visual Basic	Go, Java, Ruby, Node. js	Node.js, Python, Swift, PHP
Support for Custom Runtimes	Yes, using custom deployment packages or AWS Lambda Layers	Yes, using Azure Functions custom handlers	Yes, using custom Docker images	custom Dockerfiles	Docker skeleton
Concurrency limit	Standard: 1000 per region (soft limit) Reserved: varies Provisioned: varies	No advertised concurrency limit	No advertised concurrency limit		Standard: 1000
Resource Memory	128 M–10240 MB	128 MB–1500 MB (Consumption Plan) 128 MB–14000 MB (Premium and Dedicated Plans)	128 MB–4096 MB (in multiples of 128 MB)	128 MB–1024 MB	128 MB–2048 MB
HTTP Integration Support	Yes	Yes	Yes		Yes
Granular IM cost	1M requests per month free, then $0.20/1M requests, plus $0.00001667/GB-sec	1M requests per month free then $0.20/1M executions, plus $0.000016/GB-sec	2M requests per month free, then $0.40/1M invocations, plus $0.0000165/GB-sec	2M requests per month free, then $0.0000002/2M invocation, plus $0.00001417/GB-sec	5M requests for free, then $0.000017/GB-sec.

(Continued)

TABLE 8.1 (CONTINUED)

Serverless Services Comparisons (Tropeano, 2019; Harvey 2020; Rifai 2021; Butusov, 2021; Heller, 2021; IBM Cloud, 2021; IBM, 2021a; EOCF, 2021; OFF, 2021)

Services and Product Name	Amazon	Microsoft	Google	Oracle	IBM
Examples	Netflix, The Seattle Times, Financial Engines	FUJIFILM, Relativity	Lucille Games, HomeAway		
Pros	• Dominant market position • Extensive, mature offerings • Support for large organizations • Extensive training • Global reach • Can build app by using Lambda with other web services. • Scalability	• Second largest provider • Integration with Microsoft tools and software • Broad feature set • Hybrid cloud • Support for open source • Great debugging support • Instant deployment	• Designed for cloud-native businesses • Commitment to open source and portability • Deep discounts and flexible contracts • DevOps expertise • automatic scalability		
Cons	• Difficult to use • Cost management • Overwhelming options • Security Issues	• Issues with documentation • Incomplete management tooling • Technical support	• Late entrant to IaaS market • Fewer features and services • Historically not as enterprise-focused • Network performance		

Advantages of Serverless IoT Systems

(See Leitersdorf et al., 2017; Malawski et al., 2017;DdoS, 2020.)

1. **Scalability:** as an essential as well as robust property of serverless architecture, it allows the user to increase the server space with specific nominal costs.
2. **Big data generation:** as IoT sensors generate continuous data, more space and processing of data are required. This can be handled easily through a serverless system.
3. **Billing model:** serverless architecture has crucial advantages for the billing model, it is reducing the costs which is making it more demanding and adaptable, and working on the true pay-per-use model. The customer only has to pay for the computational resources that the function has executed. That means the customer will not be charged by the company if other functions have not been executed.
4. **Security:** serverless architecture cares about security also. At the initial stage, every small company has a lack of management tools and security experts. And if they use a server, it will be more expensive.

Due to the advantages of serverless IoT, dedicated servers are generally not preferred these days because with serverless systems, there is no need to purchase hardware and operating system licensing. Users don't need to worry about such responsibilities, as the third-party provider that sets up the server is actually responsible for maintaining that server and licensing the operating system. These service providers also manage the operating system. Serverless systems make it possible for developers to use servers and resources such as memory and computing power based on pay-per-use. This means the user only has to pay for the time that the service is availed of, which is cost effective and less expensive than using dedicated servers. Also, in serverless systems, there is no need to hire a person to maintain the servers and their runtime failure problems. One of the most important aspects is security. With a physically set up server, connected to the internet, security is a critical concern and a big challenge for users. But with serverless systems, developers can depend on a highly secure service provided by a third party and can therefore concentrate on their businesses.

8.5 Conclusions

Overcoming hazardous issues regarding public health and the environment is risky as well as time consuming. Thus, recent developments like IoT have played a pivotal role in addressing such issues. Further, running these IoT-based smartly designed waste management systems on serverless IoT architecture is a more appropriate mechanism because of resource constraints, it being easy to learn and operate, it is highly secure, and allows a continuous flow of data. There are many companies who provide this service at a reasonable cost. It can also be concluded that use of serverless systems for data storage and backend operations helps to resolve issues related to skilled labor availability.

References

Abidi, F. and Singh, V. 2013. Cloud servers vs. dedicated servers: A survey. In 2013 IEEE International Conference in MOOC, Innovation and Technology in Education (MITE) (pp. 1–5). IEEE.

Aman, M. N. Sikdar, B. Chua, K. C. and Ali, A. 2018. Low power data integrity in IoT systems. *IEEE Internet of Things Journal*, 5(4), 3102–3113.

Baldini, I., Castro, P., Chang, K., Cheng, P., Fink, S., Ishakian, V., Mitchell, N., Muthusamy, V., Rabbah, R., Slominski, A. and Suter, P. 2017. *Serverless Computing: Current Trends and Open Problems*. Cham: Springer Nature Singapore Pte Ltd.

BEHRTECH. 2020. Top 10 IoT sensor types. Available at https://behrtech.com/blog/top-10-iot-sensor-types/

Bine. 2020. The new bin-e the brand-new version of the worldwide smartest waste bin. Available at https://www.bine.world

Brous, P. Janssen, M. and Herder, P. 2020. The dual effects of the Internet of Things (IoT): A systematic review of the benefits and risks of IoT adoption by organizations. *International Journal of Information Management*, 51, 101952.

Butusov, M. 2021. AWS Lambda vs Google Cloud Functions vs Azure Functions: What to choose in 2021? Available at https://blog.techmagic.co/aws-lambda-vs-google-cloud-functions-vs-azure-functions-what-to-choose-in-2020/ (Accessed on April 21, 2021).

Central Public Health and Environmental Engineering Organisation (CPHEEO). 2018. Guidelines on usage of refuse derived fuel in various industries. Available at http://cpheeo.gov.in/upload/5bda791e5afb3SBMRDFBook.pdf

CH. 2021. The Contelligent Hub Hear from our experts and stay up to date on the latest in closed container monitoring and analytics. Available at https://contelligent.com/oneplus-news-events/oneplus-partner-program-for-waste-recycling-service-providers/

Chen, X. Y. and Jin, Z. G. 2012. Research on key technology and applications for internet of things. *Physics Procedia*, 33, 561–566.

CPCB. 2018. Solid waste generation in metrocities, India. Available at https://cpcb.nic.in/uploads/MSW/trend_46_cities_list.pdf

David, J. Langley, J. Doorn, J.V. Irene, C.L.Ng. Stieglitz, S. Lazovik, A. and Boonstra A. 2019. The Internet of Everything: Smart things and their impact on business models. *Journal of Business Research*. https://doi.org/10.1016/j.jbusres.2019.12.035

DDoS. 2020. What is a DDoS attack? Available at https://www.digitalattackmap.com/understanding-ddos/ (Accessed on 28 April 2020).

EOCF. 2021. Explore oracle cloud function. Available at https://www.oracle.com/in/cloud-native/functions/ (Accessed on May 2021).

Esquer, I. J. E. González-Navarro, F. F. Flores-Rios, B. L. Burtseva, L. and Astorga-Vargas, M. A. 2017. Tracking the evolution of the internet of things concept across different application domains. *Sensors*, 17(6), 1379.

Gramer, R. 2016. India overtakes Britain as the world's sixth-largest economy. Available at https://foreignpolicy.com/2016/12/20/india-overtakes-britain-as-the-worlds-sixth-largest-economy/ (Accessed on 05 June 2021).

Harvey, C. 2020. AWS vs Azure vs Google Cloud: 2021 Cloud Platform Comparison. Available at https://www.datamation.com/cloud/aws-vs-azure-vs-google-cloud/ (Accessed on April 22, 2021).

Heller, M. 2021. How to choose a cloud serverless platform. Available at https://www.infoworld.com/article/3605129/how-to-choose-a-cloud-serverless-platform.html (Accessed on March 3, 2021).

IBM. 2021a. IBM cloud docs. Available at https://cloud.ibm.com/docs/openwhisk?topic= openwhisk-runtimes (Accessed on May 2021).

IBM. 2021b. Serverless in the enterprise, 2021: Building the next generation of efficient, flexible, cost-effective cloud native applications. Available at https://www.ibm.com/downloads/cas/ZJLWQOAQ (Accessed on 5 June 2021).

IBM Cloud. 2021. https://cloud.ibm.com/functions/learn/pricin (Accessed on 5 June 2021).

Keramidas, G. Voros, N. and Hübner, M. 2016. *Components and Services for IoT Platforms*. Cham: Springer International Pu.

Khvoynitskaya, S. 2019. The IoT history and future. Available at https://www.itransition.com/blog/iot-history

Leitersdorf, Y. Schreiber, O. Reznikov, I. and Ninyo, I. 2017. The big opportunities in serverless computing. Available at https://venturebeat.com/2017/10/22/the-big-opportunities-in-serverless-computing/ (Accessed on 18 March 2021).

Malawski, M., Gajek, A., Zima, A., Balis, B. and Figiela, K. 2017. Serverless execution of scientific workflows: Experiments with HyperFlow, AWS Lambda and Google Cloud Functions. *Future Generation Computer Systems*. https://doi.org/10.1016/j.future.2017.10.029

Misra, D. Das, G. Chakrabortty, T. and Das, D. 2018. IoT-based waste management system monitored by cloud. *Journal of Material Cycles and Waste Management* 20, 1574–1582.

OFF. 2021. Oracle Function FAQ. Available at https://www.oracle.com/webfolder/technetwork/tutorials/FAQs/oci/Functions-FAQ.pdf (Accessed on 20 April 2021).

Palanivel, T. M. and Sulaiman, H. 2014. Generation and composition of municipal solid waste (MSW) in Muscat, Sultanate of Oman. *APCBEE Procedia*, 10, 96–102.

Patra, S. S., and Barik, R. K. 2015. Dynamic dedicated server allocation for service oriented multi-agent data intensive architecture in biomedical and geospatial cloud. In *Cloud technology: Concepts, methodologies, tools, and applications* (pp. 2262–2273). IGI Global.

Paulraj, C. R. K. J., Bernard, M. A., Raju, J. and Abdulmajid, M. 2019. Sustainable waste management through waste to energy technologies in India-opportunities and environmental impacts. *International Journal of Renewable Energy Research (IJRER)*, 9(1), 309–342.

Rifai M. 2021. A cloud Guru. Lambda vs Azure Functions vs Google Cloud Functions. Available at https://acloudguru.com/blog/engineering/serverless-showdown-aws-lambda-vs-azure-functions-vs-google-cloud-functions (Accessed on April 29, 2021).

SB. 2021. Smart solutions for cities. Available at https://www.cleanindiajournal.com/sensor-based-smart-bins/

Sethi, P. and Sarangi, S. R. 2017. Internet of things: Architectures, protocols, and applications. *Journal of Electrical and Computer Engineering*, 2017, 1–25. https://www.hindawi.com/journals/jece/2017/9324035/.

Sharma, K.D. and Jain, S. 2020. Municipal solid waste generation, composition, and management: the global scenario. *Social Responsibility Journal*, 16(6), 917–948.

Shekdar, A. V. 2009. Sustainable solid waste management: an integrated approach for Asian countries. *Waste Management*, 29(4), 1438–1448.

Todd, D. 2019. Understanding edge architecture through the IIoT lens. Marketing communications specialist, stratus technologies. Available at https://blog.stratus.com/understanding-edge-architecture-through-iiot-lens/

Tropeano, D. 2019. IBM Cloud Functions Raises Operations Limits. Available at https://www.ibm.com/cloud/blog/announcements/ibm-cloud-functions-raises-operations-limits (Accessed on 25 May 2021).

Truong, H. L. and Dustdar, S. 2015. Principles for engineering IoT cloud systems. *IEEE Cloud Computing*, 2(2), 68–76.

UNEP. 2015. United Nations Environment Programme. Global Waste Management Outlook. Available at https://www.uncclearn.org/wp-content/uploads/library/unep23092015.pdf

Wang, H. Niu, D. and Li, B. 2019. Distributed machine learning with a serverless architecture. In IEEE INFOCOM 2019-IEEE Conference on Computer Communications (pp. 1288–1296).

World Bank. 2018. World Bank: Global waste generation. Available at https://www.wastedive.com/news/world-bank-global-waste-generation-2050/533031/ (Accessed on April 30, 2021).

9

IoT in Hospital Solid Waste Generation and Management

Uma Rahangdale, Amar Shinde, Gazala Yasmin Ashraf, and Vipin Singh

CONTENTS

9.1 Introduction

The hospital and healthcare industry has become one of the fastest-growing sectors in past decades (Yeoh et al., 2013). The infrastructure of the healthcare industry is increasing with the increase in population, further leading to an increment in budget and investment are for this industry. It was estimated that around 7.8 trillion USD worldwide for the year 2017 was spent on the infrastructure of the healthcare industry, which is nearly 10% of global GDP. It is further estimated to be approximately 8.6 billion USD in 2030 and 9.8 billion USD in 2050. Healthcare global market opportunities and strategies to 2022 reported that the rapidly growing healthcare industry reached a value of nearly $8,452 billion in 2018 with a compound annual growth rate (CAGR) of 7.3%. It is expected to further grow up to $11,909 billion by 2022. Moreover, the healthcare industry in India is growing at a rate of 17% annually and is expected to reach revenues of $300 billion by 2022 (BW, 2019). The global Internet of Things (IoT) in hospitals and healthcare is also growing simultaneously. Business of IoT in this sector is expected to increase from ~72.5 billion (USD) in the year 2020 to ~188.2 billion (USD) by the year 2025, at a CAGR of ~21%. The growth factors for IoT in this sector are the increasing focus on active patient engagement and patient-centric care, monitoring measures for cost control, development of high-speed networking technologies connectivity, and patient-centric service delivery through various possible ways (IHM, 2020). The increase in the healthcare facilities for patient treatment will result in the production of large amounts of healthcare waste (Wilson et al., 2015; BW, 2019; GS, 2019). According to CPCB (2018), there are about 270,416 healthcare facilities in India, where 97,382 are bedded facilities with 2,206,362 beds. An approximately 44% increase in healthcare facilities was reported between 2016 and 2018. Healthcare waste is being produced in hospitals and needs to be collected, transferred, and disposed of. The Indian Government initiated the first set of rules for biomedical waste management in 1998 via the Ministry of Environment and Forest (MOEF). These rules were set out to deal with the definition and segregation categories of biomedical waste, and in order to protect those who handle the waste, as well as its treatment, etc. A recent amendment for biomedical waste management came in the year 2016, which further simplified categorization, authorization, and improvement in different steps of healthcare waste handling (Shettennavar and Vithayathil, 2019).

9.2 Healthcare Waste (HCW) Generation, Exposure, and Impact

According to WHO (1985), healthcare waste (HCW) or biomedical waste (BMW) includes all the waste generated from medical procedures within healthcare facilities (HCFs). Sources of hospital waste are laboratory cultures waste, and sharps objects like disposable needles, syringes, etc. Pathological waste includes tissues, human flesh, blood and body fluids, and pharmaceuticals drugs and chemicals treatment (WHO, 2005).

Nemathaga et al. (2008) investigated healthcare waste generation and its management practice in South Africa. The study was done in two hospitals. Firstly, Tshilidzini hospital serves a town population of nearly 26,000. This contains 530 beds (450 patients were admitted during this study visit) with an average of 160 admissions per day and 340 in outpatient (OPD) attendance. There are about 1,800 members of staff in the hospital. Ten major wards are in this hospital. The second hospital was Elim hospital, which has about 1,000 members of staff. There are a total of 323 beds in this hospital, where average patient admission was 70 per day (250 admitted during this study visit). The hospital includes nine major wards.

The study of Nemathaga et al. (2008) reported generation of HCW was found to be 0.55 kg/patient for Tshilidzini, and 0.65 kg/patient for Elim hospital. Total daily HCW production in Tshilidzini was found to be ~350 kg/day, which included ~225 kg general waste, 25 kg sharp objects, and ~100 kg medical waste, whereas in Elim hospital, it was about 229 kg/day, consisting of 131 kg general waste, ~25 kg sharp objects, and 73 kg medical waste (Figure 9.1). The HCW by ward (including kitchen) was further investigated in the study, where it was found that the maternity ward generated the highest amount of waste in both hospitals (Figure 9.2). HCW may have serious impacts on health and environment, and it was also reported that exposure to HCW may have serious impacts on human health such as mutagenic and carcinogenic effects, effects on the respiratory system, and nervous and reproductive systems.

Hasan and Rahman (2018) studied healthcare waste management paradigms in Khulna, the southwestern division of Bangladesh. The study was done by surveying 20 different hospitals including diagnostic centers, a total of ~3000 people.

Overall BMW generation rate and hazardous BMW generation rate were 0.90 kg/bed/day and 0.18 kg/bed/day. An assessment of the management systems revealed that ~56% of workers did not receive any form of training in the handling of hazardous waste. Around 54% of them did not use any safety equipment or clothing. This study reported that a major part of hospital waste is non-hazardous in nature. Approximately 20% is hazardous, which further includes 8% of infectious waste, 1% chemicals, 2% sharp objects, and 5% pathological waste (Figure 9.3). Kenny and Priyadarshini (2021) reported per bed hospital waste generation for various countries, according to which developing countries generate a high level of HCW. It was observed that developed countries like the United States, Canada, and Ireland generate

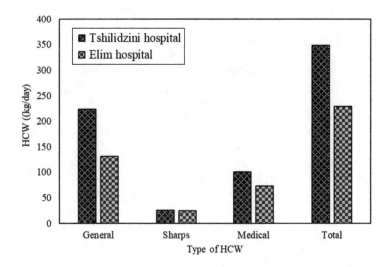

FIGURE 9.1
Type of healthcare waste generation in two hospitals of South Africa (Nemathaga et al., 2008).

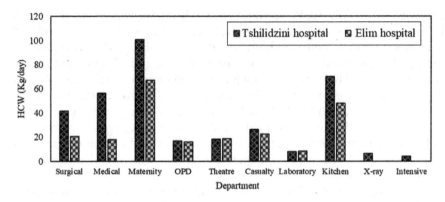

FIGURE 9.2
Department-wise healthcare waste generation in two hospitals of South Africa (Nemathaga et al., 2008).

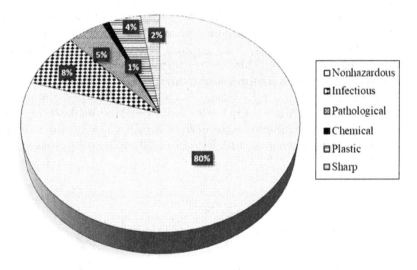

FIGURE 9.3
Composition of hospital waste (Hasan and Rahman, 2018).

more hospital waste. It was found to be the highest i.e., 8.4 kg/bed/day in the United States, and lowest for Nepal i.e., 0.5 kg/bed/day. For India, the average rate of HCW was found to be 1.6 kg/bed/day (Figure 9.4).

Yenesew et al. (2012) did a study for risk factors associated with healthcare staff during handling of HCW in a health care center located in Gondar Town in Ethiopia. It was performed during April–May 2011, where about 260 healthcare members of staff were included and random sampling was done for the collection of data and information. The study found about 60% of healthcare staff members had low risk perception illustrated by improper handling of HCW. There is lack of awareness and training for the waste

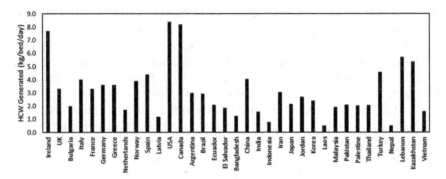

FIGURE 9.4
Healthcare waste generation rate worldwide (Kenny and Priyadarshini, 2021).

management plan and the guidelines defined therein. Tomkins et al. (2012) further reported occupational transmission of hepatitis C (HCV) in hospital staff and associated risk factors in the United Kingdom. Ten years of exposure data were analyzed for the duration of 1997 to 2007. Approximately 15 cases were documented that occurred due to percutaneous exposure from hollow needles that were contaminated with blood fluid. Sharma et al. (2013) aimed to investigate the awareness of HCW among health care staff in India. The cross-sectional investigation was conducted with 144 staff members, i.e., doctors, nurses, technicians, cleaners, and maintenance workers in a dental college and hospital. The study documented that about 50% of housekeepers and maintenance workers reported at least one incident with exposure to HCW in a year. Approximately 36% of nurses had extremely poor knowledge about HCW management practice.

According to Asante et al. (2014), HCW is an issue for public health safety as well as being of environmental concern because of its hazardous and infectious nature. HCW is a recent major concern for every nation including developing countries like (Greater Accra Region) Ghana, where there is minimal knowledge and information about the generation of HCW and how to manage and dispose of it. The study was conducted over more than 100 healthcare centers, where about 150 healthcare officers and workers responsible for waste management were included. The study found that ~8221 kg HCW was produced for a total of 6,851 beds in the region. There is a lack of regulatory policy in the region, which is a current matter of consideration by local government in order to implement further recommendations. This study also found that as HCW is increasing greatly, there is a requirement for a strategic management focus at a priority level that should be at international standards. Dwivedi et al. (2009) reported that mainly big hospitals in India had safe disposal of waste, whereas there were no proper facilities for HCW handling in small healthcare centers and even public hospitals. Another study by Vijaya et al. (2007) reported that HCW storage and segregation bins are (a) far away from nursing stations as well as from patients; (b) these are uncovered; because of which flies, dogs, and rats could spread the waste and infection; (c) as these bins are not placed in a specific designated location for storage, random dumping of HCW was observed; (d) delays in transport of HCW was observed; as per guidelines, bins should be emptied or HCW transported to central facilities within 24 hours. The irregularity of transportation of HCW was also reported in another study done by Verma (2010). The study suggested that infected waste should be transferred with high level of precaution and care, with separate air-conditioned chambers.

Mohankumar and Kottaiveeran (2011) reported a rise of environmental issues due to HCW (e.g., pathological waste), which is increasing regularly in India. The study reported there are about 38,000 small, private, and primary health centers. HCW management costs about 8% of complete management of waste in India and may further increase by ~20%. The study reported that regulation for biomedical waste is made and is strict towards its implementation, but still it could be observed that many institutions, dispensaries, clinics, chemist shops, and hospitals ignore proper handling of HCW. Such mishandling causes exposure to infection and threats to the environment. Shivalli and Sanklapur (2014) studied nurses' roles in healthcare waste management. The study was conducted in a tertiary care hospital in Mangalore city in India. The hospital had a capacity of >600 beds and generated all types of medical waste. The total of working nurses in the hospital was 198. A total of 100 nurses participated in the study, and the majority (82%) of them were female. The mean age of the nurses was 26. A random survey was done. The analysis was done using the Statistical Package for the Social Sciences (SPSS), which shows nurses' attitudes towards healthcare waste management. Out of 100 nurses, 47 had excellent knowledge (>70% score). Most (86%) expressed the need for refresher training. Knowledge of waste management practices was not satisfactory. There is a need for refresher training at optimum intervals to ensure sustainability and further improvement that could be provided in local language. Another study done by Das and Biswas (2016) also reported that healthcare waste produced during treatment in hospitals is hazardous. It can be injurious to the environment and health. Appropriate management of HCW is the responsibility of every citizen. This study was conducted in order to investigate awareness and observe the practice of HCW management in West Bengal, India in healthcare centers. This study was conducted among various healthcare workers e.g., doctors of various departments, radiologists, and other staff. It revealed there is certain lack of awareness among healthcare providers, and there is a need for regular comprehensive training for health staff. Letho et al. (2021) studied practice and awareness HCW management in National Referral Hospitals in Bhutan for March–April 2019 using various methods such as a demographic questionnaire, awareness questions, and an observational checklist. Their statistical results showed that most respondents in this study were female (~54%) and their average age was ~32. It was observed that about ~57% of them did not receive any kind of training or education in HCW management, whereas

~74% were aware of HCW and ~98% for personal protection. It was reported that only ~38% were aware that HCW can only be kept on a hospital site for 48 hours, and ~61% observed correct segregation according to national HCW guidelines. Out of this, about 50% of the waste was not transported properly. Lima et al. (2017) further reported that for healthcare workers, biological waste exposure is commonly observed. About 86% of nursing staff and technicians were affected through this exposure, whereas 83% was due to exposure to sharp healthcare waste materials in Brazil.

9.3 Traditional Approach for Healthcare Waste Management

Healthcare waste is collected and segregated manually based on color-coded waste containers or bins (Figure 9.5). There are four types of containers: (a) yellow bag for collection of human anatomical waste, (b) red bag for collection of infectious waste, (c) blue bag for plastic and glass waste, and (d) white bag or puncture-proof bag for sharp waste (Rao, 2009). This waste is transported by vans to a central facility for waste treatment i.e., a common waste treatment facility for medical or healthcare waste treatment. These common treatment facilities receive a huge quantity of healthcare waste which is further quantified manually, helping to maintain records as well as proper pretreatment. The governing body regularly monitors these facilities for their appropriate functionality. It demands annual reports from the central facilities to maintain records and to analyze data (Raundale et al. 2017).

Biomedical waste should be disposed of with utmost precaution, as some waste can be contaminated with microorganisms. The different disposal practices are deployed based on the level of contamination (HWE, 2013;

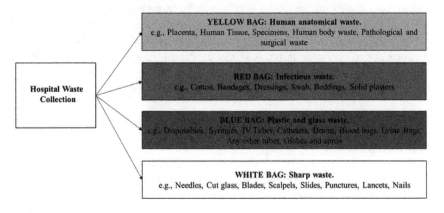

FIGURE 9.5
Types of container and color coding for hospital waste collection (Rao, 2009).

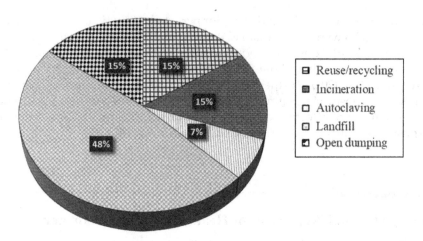

FIGURE 9.6
Method of healthcare waste handling (Nemathaga et al., 2008).

Voudrias, 2016). (a) Autoclaving: highly contaminated waste should be autoclaved for the complete destruction of microorganisms before being taken to the landfill. (b) Incineration: a very large portion of the waste is generally incinerated, as it is a quick, economical, and easy method to dispose waste, with huge volume reduction. This process occurs at high temperature (82°C–1,093°C), thus destroying pathogens.

However, the after-treatment devices, viz. wet scrubber and electrostatic precipitator, should be deployed to remove the emissions produced during burning the waste. (c) Chemical disinfection: this is commonly adopted for the sanitization of liquid biomedical waste. Chemical disinfection can also be used for the solid waste but only after grinding the solids in order to ensure maximum decontamination. Some recent studies have recommended microwaving over incineration of the waste, as it is more efficient and profoundly lowers the volume of the waste. During the microwaving process, the water is mixed into shredded waste and the mixture is heated for the decontamination of the microorganisms. A study by Nemathaga et al. (2008) for HCW of hospitals reported landfill (~50%) was common practice in South Africa, whereas incineration (~15%) and recycling (~15%) were also seen (Figure 9.6).

9.4 Limitations of Traditional Approach for Healthcare Waste Management

Although this system has been working well for decades, there are certain flaws which are potentially hazardous to human health as well as to the environment. There are some other limitations associated with this system: (a) weak decision making; (b) possibility of human error at each step, as quantification and segregation are done by staff; (c) chances of fraud because some hospitals could hide the facts; (d) delays in the collection of data, as it is done by humans; (e) corruption can occur; (f) not real time, so it is not possible to get instant and actual data for a plan of action. Nowadays, to address these limitations of various sectors, there is an advanced and real-time solution that is based on the Internet of Things (IoT).

9.5 IoT in Healthcare Waste Management

9.5.1 Definition and Application of IoT

The Internet of Things (IoT) is a revolutionary development that helps in communication between devices based on electronics and sensors via the internet, making things easier in our daily lives thanks to these industrial applications. In order to give an innovative solution, there is a use of smart equipment in IoT. This solution could fulfill the requirements of different businesses, industries, or government organizations (Kumar et al., 2019).

Nowadays, IoT is becoming an important feature of our lives and can be found around us, consisting of a smart framework, architecture, intelligent equipment, and dedicated sensors. Based on 640 actual IoT projects around the globe, the top areas of IoT application were found (Figure 9.7) to be industry, smart city projects, smart energy, connected vehicles, smart agriculture, connected buildings, connected healthcare, and retail and supply chains (IoT

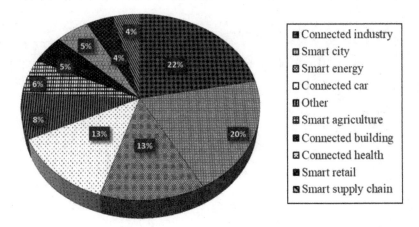

FIGURE 9.7
The top ten IoT application areas based on real IoT projects (IoT AA, 2016).

AA, 2016). A high number of projects in the United States (~44%) are based on IoT, followed by the European Union with ~34%. The variation is observed in individual applications of IoT in the regions. North America, specifically the United States, is very strong in connected health as well as smart retail i.e., about 61% and 52% respectively. On the other hand, in Europe the focus is mainly on smart city projects i.e., about 47%. Asian regions use IoT mainly in smart energy protection i.e., about 25% (IoT AA, 2016; Kumar et al., 2019).

With an increase in globalization around the world, there is an increase in all types of waste, giving serious issues to each governing body, stakeholders, etc. The world's urban population will reach 4.3 billion in 2025 as per the estimation of the World Bank, which will generate approximately 1.42 kg/capita/day. Similarly, there will be a rise in healthcare waste as well. At the moment, IoT-based smart waste management is functional in many sectors in different parts of the world for their waste management systems, e.g., in industry, healthcare, handling of domestic waste, etc. (Ismail et al., 2019). In other words, due to increases in population and rapid urbanization, of great importance in every country is focusing on good health, quality of environment, and maintaining hygiene. It has been observed that solid waste management is one of the emerging challenges for all of us and creates environmental as well as health issues. Several steps and initiatives have been taken towards a sustainable environment by using smart solutions that could minimize negative impacts. IoT is one of those initiatives that is applied in the area of solid waste management.

9.5.2 How IoT Works for Waste Management in Healthcare Systems

The framework of IoT-based healthcare waste (HCW) management contains an overview of the different components associated with it (Figure 9.8): (a) the sensors are hardware that is required for the measurement of waste in

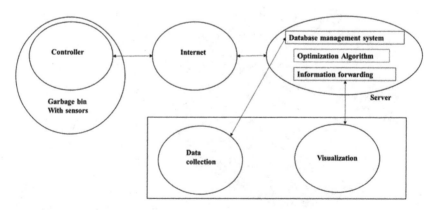

FIGURE 9.8
Framework for IoT-based HCW management system (Shyam et al., 2017).

particular garbage bins, which is based on sonar, e.g., Ultrasonic Ranging Module (HC-SR04); (b) the network interface is the part of data collection unit, where it is further sent to the remote server using internet or wireless networks; (c) database systems, for example MySql, which are used for storage of all data collected by the sensors and the trucks; (d) artificial intelligence (AI) used for forecasting level/type of waste; (e) optimization algorithms that help to identify which wastebins have been emptied, and also that the shortest path for collection is taken; (f) information adaptation and forwarding step helps in the destination path and must be sent to the collectors in understandable format (Shyam et al., 2017).

The components of an IoT devices-based smart healthcare waste management system (Kumar et al., 2016; Raundale et al., 2017; Srikanth et al., 2019).

(a) Smart color-coded garbage bins
(b) Microcomputer/Raspberry Pi/Arduino Uno microcontroller/RFID
(c) Application server
(d) Notification server
(e) Local database
(f) Self-identifying disposable baggage
(g) EC05 Bluetooth module
(h) Power Supply – 230V AC changed to 12V, 5V DC
(i) HX711 Strain gauge drivers (x2)
(j) L393 comparator
(k) Smoke sensor

The expenses of IoT-based systems are mainly expressed in terms of the cost of hardware. According to Raundale et al. (2017), it is $5 for an IoT microcomputer, $1 for weight sensors, $1 for tilt sensor, $15 for database/central servers, $5 for local servers, and $2 for Radio-frequency identification (RFID) tags/reader.

The IoT system developed using recent technological advancements and internet connectivity helps to automate the waste management process of the healthcare system. Automating the system using IoT devices makes steps easy, which is also affordable and easy to install. Commercial off-the-shelf (COTS) components are commonly used for IoT. In this IoT-based smart HCW management system, existing internet facilities can be used for the transfer of real-time data. The system is generally enabled and implemented at the source to the existing color-coded smart waste bins where RFID tags are be installed that automatically index using a system. These smart waste bins are further installed with weighing sensors, which immediately measure weight when waste is put into it. This measured value of each color-coded bin is sent by microcomputer using the internet to the server or person authorized. These records can be checked at any time, as they are based on real-time measurement. The instant transfer of information helps to reduce human error as well as ensure the correct information. This system is entirely distributed and based on real-time data; thus, it can help government agencies and municipal corporations to analyze the data and take appropriate actions. Some other techniques like big data could be further useful for the classification of collected data. The collected data can be stored in centralized or decentralized servers, based on the requirement (Kumar et al., 2016; Raundale et al., 2017; Srikanth et al., 2019). An Indian study by Gade and Aithal (2021) looked at information and communications technology (ICT) and IoT-based solutions for managing waste smartly. The device was named ismartWMS. It was built with IoT sensors along with a cloud server that provided a graphical user interface. Pardini et al. (2019) suggested in their study that there are tremendous issues related to the production and handling of waste, including HCW. The role of automated systems using IoT and cloud computing is changing the performance of waste management.

9.5.3 Advantages of IoT-Based Healthcare Waste Management over Traditional Approach

As discussed earlier, the traditional system of HCW management is monitored manually, and in many cases, municipal workers do not know if the garbage bin is full or empty. Sometimes due to excess fill, the garbage spills over around and causes inconvenience to people, as they are exposed to bad odors as well as infections. In addition, when we see automated IoT-based HCW management systems, they have multiple benefits as follows (Malini and Hemalatha, 2019).

The HCW gets stored in color-coded garbage bins with self-identified objects and QR code marked. This helps in contactless source segregation.

(a) It is well developed and integrated through an internet network, making it easy to share information/data of waste produced at any time, as it is based on real-time measurement

(b) These smart garbage bins could be monitored automatically

(c) The bin senses when it is raining and closes the door/cover of the garbage bin automatically, helping to restrict leachate or prevent the spread of odors around it

(d) Alarm indicator to the authorities when the garbage bin is filled

(e) Fast and well-organized database because it is a human-less system, easy to analyze

(f) Timely and fast collection and transport of HCW to the treatment site

(g) As it is an automated system, there are no or fewer chances of exposure to disease that leads to occupational safety of staff

(h) Helps to reduce the chances of fraud/mismanagement or manipulation of records

9.6 Conclusions

The healthcare system is of high priority in the world. It is strongly influenced by economic circumstances as well. Large healthcare infrastructures and treatment facilities produce a huge quantity of healthcare waste. Its proper management is of vital importance at the moment. Technological development occurs in every sector of our lives. Thus, a developed automated system or IoT-based system offers an advanced and affordable healthcare waste management system and plays a pivotal role in managing it intelligently, in a timely manner, and with minimal risks to human health or the environment.

References

Asante, B. Yanful, E. and Yaokumah, B. 2014. Healthcare waste management; its impact: A case study of The Greater Accra Region, Ghana. *International Journal of Scientific & Technology Research*, 3(3), 106–112.

BW 2019. The $11.9 trillion global healthcare market: Key opportunities & strategies (2014–2022). Available at https://www.businesswire.com/news/home /20190625005862/ en/ The-11.9-Trillion-Global-Healthcare-Market-Key-Opp ortunities-Strategies-2014-2022---ResearchAndMarkets.com

CPCB 2018. Annual report on biomedical waste management. Available at https:// cpcb.nic.in/uploads/Projects/Bio-Medical-Waste/AR_BMWM_2018.pdf

Das, S. K. and Biswas, R. 2016. Awareness and practice of biomedical waste management among healthcare providers in a Tertiary Care Hospital of West Bengal, India. *International Journal of Medicine and Public Health*, 6(1), 19–25. DOI: 10.4103/2230-8598.179755.

Dwivedi, A. K. Pandey, S. and Shashi 2009. Fate of hospital waste in India. *Biology and Medicine*, 1(3), 25–32.

Gade, D. S. and Aithal, P. S. 2021. Smart city waste management through ICT and IoT driven Solution. *International Journal of Applied Engineering and Management Letters (IJAEML)*, 5(1), 51–65.

GS 2019. Global spending on health: A World in transition-WHO. Available online: https://www.who.int/health_financing/documents/health-expenditure -report-2019.pdf?ua=1 (accessed on 10 May 2021).

Hasan, M. M. and Rahman, M. H. 2018. Assessment of healthcare waste management paradigms and its suitable treatment alternative: A case study. *Journal of Environmental and Public Health*, 2018, 1–14. DOI: 10.1155/2018/6879751.

HWE 2013. Hazardous waste experts. Regulated medical waste treatment methods. Available at https://www.hazardouswasteexperts.com/regulated-medical -waste-treatment-methods (Accessed on 28 May 2021).

IHM 2020. IoT in healthcare market. Available at https://www.marketsandmarkets .com /Market-Reports/iot-healthcare-market-160082804.html.

IoT AA 2016. IoT application areas. https ://iot-analy tics.com/top-10-iot-proje ct-appli catio n-areas -q3-2016/ (Accessed 05 June 2021).

Ismail, N. A. Ab Majid, N. A. and Hassan, S. A. 2019. IoT-based smart solid waste management system: A systematic literature review. *International Journal of Innovative Technology and Exploring Engineering (IJITEE)*, 8(8), 1456–1462.

Kenny, C. and Priyadarshini, A. 2021. Review of current healthcare waste management methods and their effect on global health. In *Healthcare* (Vol. 9, No. 3, p. 284). Multidisciplinary Digital Publishing Institute.

Kumar, N. S. Vuayalakshmi, B. Prarthana, R. J. and Shankar, A. 2016. IoT based smart garbage alert system using Arduino UNO. In 2016 IEEE Region 10 Conference (TENCON), Singapore (pp. 1028–1034). IEEE.

Kumar, S. Tiwari, P. and Zymbler, M. 2019. Internet of Things is a revolutionary approach for future technology enhancement: A review. *Journal of Big Data*, 6(1), 1–21.

Letho, Z. Yangdon, T. Lhamo, C. Limbu, C. B. Yoezer, S. Jamtsho, T. and Tshering, D. 2021. Awareness and practice of medical waste management among healthcare providers in National Referral Hospital. *Plos One*, 16(1), 1–10. DOI: 10.1371/journal.pone.0243817.

Lima, G. M. N. Kawanami, G. H. and Romeiro, F. G. 2017. Occupational exposures to biological material among health professionals of Bauru Base Hospital: Preventive and post-exposure measures. *Revista Brasileira de Medicina do Trabalho*, 15(3), 194.

Malini, V. and Hemalatha, M. 2019. IoT based smart garbage alert system using wireless sensor network for environmental hygiene. *International Journal of Engineering and Advanced Technology (IJEAT)*, 8(6), 3601–3603. DOI: 10.35940/ijeat.F9356.088619.

Mohankumar, S. and Kottaiveeran, K. 2011. Hospital waste management and environmental problems in India. *International Journal of Pharmaceutical & Biological Archives* 2(6), 1621–1626.

Nemathaga, F. Maringa, S. and Chimuka, L. 2008. Hospital solid waste management practices in Limpopo Province, South Africa: A case study of two hospitals. *Waste Management*, 28(7), 1236–1245.

Pardini, K. Rodrigues, J. J. Kozlov, S. A. Kumar, N. and Furtado, V. 2019. IoT-based solid waste management solutions: A survey. *Journal of Sensor and Actuator Networks*, 8(1), 5.

Rao, Hanumantha P. 2009. Hospital waste management system: A case study of a south Indian city. *Waste Management & Research*, 27(4), 313–321.

Raundale, P. Gadagi, S. and Acharya, C. 2017. IoT based biomedical waste classification, quantification and management. In 2017 International Conference on Computing Methodologies and Communication (ICCMC) (pp. 487–490). IEEE. Erode, India. DOI: 10.1109/ICCMC.2017.8282737.

Sharma, A. Sharma, V. Sharma, S. and Singh, P. 2013. Awareness of biomedical waste management among health care personnel in Jaipur, India. *Oral Health and Dental Management*, 12, 32–40.

Shettennavar, S. and Vithayathil, A. 2019. Exploratory study of biomedical waste management: An IoT perspective. *Asian Journal of Management*, 10(3), 181–189.

Shivalli, S. and Sanklapur, V. 2014. Healthcare waste management: Qualitative and quantitative appraisal of nurses in a tertiary care hospital of India. *The Scientific World Journal*, 2014, 1–6. http://dx.doi.org/10.1155/2014/935101.

Shyam, G. K. Manvi, S. S. and Bharti, P. 2017. Smart waste management using Internet-of-Things (IoT). In 2017 2nd International Conference on Computing and Communications Technologies (ICCCT) (pp. 199–203). IEEE. Chennai, India. DOI: 10.1109/ICCCT2.2017.7972276

Srikanth, C. S. Rayudu, T. B. Radhika, J. and Anitha, R. 2019. Smart waste management using Internet-of-Things (IoT). *International Journal of Innovative Technology and Exploring Engineering*, 8, 3–5.

Tomkins, S. E. Elford, J. and Nichols, T. 2012. Occupational transmission of hepatitis C in healthcare workers and factors associated with sero conversion: UK surveillance data. *Journal of Viral Hepatitis*, 19(3), 199–204.

Verma, L. K. 2010. Managing hospital wastes. How difficult? *Journal of Indian Society of Hospital Waste Management*, 9(1), 47–50.

Vijaya, K. G. Kavita, D. and Vidya, K. B. 2007. *A Critical Analysis of Healthcare Waste Management in Developed and Developing Countries: Case Studies from India and England* (pp. 134–141).

Voudrias, E. A. 2016. Technology selection for infectious medical waste treatment using the analytic hierarchy process. *Journal of the Air & Waste Management Association*, 66(7), 663–672.

Wilson, D. C. Rodic, L. Modak, P. Soos, R. Carpintero, A. Velis, K. and Simonett, O. 2015. *Global Waste Management Outlook*. UNEP United Nations Environment Programme. Available online: https://www.unenvironment.org/resources/report/global-wastemanagement

World Health Organization. 1985. *Word Health Organization Management of Waste from Hospitals, EURO Reports and Studies*: 97, Copenhagen.

World Health Organization (WHO). 2005. *Management of Solid Health: Care Waste at Primary Health: Care Centers, a Decision-Making Guide*. Geneva: WHO.

Yenesew, M. A. Moges, H. G. and Woldeyohannes, S. M. 2012. A cross sectional study on factors associated with risk perception of healthcare workers toward healthcare waste management in health care facilities of Gondar Town, Northwest Ethiopia. *International Journal of Infection Control*, 8(3), 1–9.

Yeoh, E. Othman, K. and Ahmad, H. 2013. Understanding medical tourists: Word-of-mouth and viral marketing as potent marketing tools. *Tourism Management*, 34, 196–201.

10

A Fast Garbage Classification Model Based on Deep Learning

Rampavan Medipelly and Earnest Paul Ijjina

CONTENTS

10.1 Introduction

Rapid growth in population and urbanization have led to a point where garbage disposal has become a major cause of concern. Throughout the world, billions of tons of waste is produced, particularly in urban areas and in highly populated areas. According to a World Bank report released in 2018, the worldwide waste generated annually across countries will increase to 3.4 billion tons in the next 30 years. By 2050, if nothing is done, the amount of garbage will increase by 70% [1].

The accumulation of huge volumes of garbage causes problems to the environment and to the community. In the places where there is no proper waste management, the garbage is either buried or burned after it is collected for disposal. If it is buried, it will pollute the soil. When it is burned, it releases toxic gases into the air, resulting in air pollution, which can lead to diseases such as cancer and asthma. When it is dumped in open spaces, it emits bad smells and spread diseases. When the rains come, it enters water resources and pollutes the water.

Garbage can be broadly classified into two groups: biodegradable and non-biodegradable. With the help of bacteria, temperature, oxygen, and other

DOI: 10.1201/9781003184096-10

factors, biodegradable waste can be decomposed and reused as manure without causing any pollution. The waste products of fruits, vegetables, paper, animals, plants etc., belong to biodegradable waste. Non-biodegradable waste, on the other hand, is any substance that cannot be decomposed easily (generally, it is not naturally available), and causes pollution by remaining in the environment for longer periods. Rubber, plastic, metal, glass, cement products, ceramic tiles, bricks, etc. are examples of non-biodegradable waste.

Most developing countries lack proper waste management systems. Garbage collection and disposal is done manually in these countries. This has many disadvantages. It is tedious, labour intensive, and harmful to the workers. The recent addition of e-waste and biomedical waste necessitated the need to investigate safe garbage management systems.

There is a need for efficient garbage management systems that can effectively recognize garbage and classify them into biodegradable and non-biodegradable garbage. Then non-biodegradable material is further divided into reusable, recyclable, combustible, and inert material, for further utilization. This will help in minimizing manual workforce, cutting expenses, increasing efficiency, and decreasing pollution. Although developed countries such as the United States, China, and countries in Europe have diverse waste management systems, there is a need for a better system. To process the enormous amounts of visual data generated from various places, the system should have a) an end-to-end model to process the large amount of visual data generated from different locations and b) real-time data storage and processing power.

The reminder of this chapter is organized as follows. Section 10.2 presents the existing related literature, Section 10.3 discusses the proposed approach. The details of the experimental study in Section 10.4. The conclusions and future work are given in Section 10.5.

10.2 Existing Work

Waste management is essentially a two-step process, where the first step is to detect and collect the garbage. The second step is sorting/classifying the waste into their respective categories for further processing. Many approaches for garbage detection have been proposed in the literature. To localize and categorize objects, Mohammad Saeed Rad et al. [2] suggested a GoogleNet [3] based architecture. Mittal et al. [4] suggested the Garbnet model, which uses Alexnet [5] as its backbone to detect and localize garbage. Bansal et al. [6] proposed a fully automatic system to detect and collect garbage. They used CNN for detecting garbage along with proportional integral and derivative control algorithms for finding the location of garbage from the camera image. Wang et al. [7] proposed a Faster RCNN [8] based garbage detection model. Zhihong et al. [9] used Region Proposal Generation (RPN) with VGG-16 model for garbage recognition. Lui et al. [10] proposed improved YOLOv2 [11] architecture for garbage detection and classification. This technique outperforms Faster RCNN-based approaches. The YOLOv3 [12] trashnet model introduced by Carolis et al. [13] detects garbage dumps and garbage bins. Panwar et al. [14] developed an aqua vision model for detecting garbage on water using deep transfer learning, which was evaluated on the AquaTrash dataset.

Those based on garbage classification differ from methods based on detection. The goal of these methods is to use image data collected from various garbage sites to separate the various classes. Kang et al. [15] used ResNet34 to classify different types of garbage and introduced multiple types of architectures for the trashnet dataset. Adedeji et al. [16] proposed waste classification system on trash dataset. They used ResNet50 pre-trained model for feature extraction and SVM for classification. Yang et al. [17] proposed a neural network-based WasNet architecture, which achieved better results on multiple garbage related datasets. Rabano et al. [18] proposed mobilenet-based architecture that is tested on trashnet dataset, which can be deployed on an Android mobile phone for detecting plastic material. Vo et al. [19] used DNN-TC, a ResNext-based model for classifying trash, which was evaluated on trashnet and VN-trashnet datasets. In Aral et al. [20] and Bircanoğlu et al. [21], several deep learning-based models such as Densenet121, DenseNet169, InceptionResnetV2, MobileNet, and Xception are evaluated on trashnet, for garbage classification.

Other waste-segregation systems based on sensors are also proposed. To separate garbage, Chandramohan et al. and Sharanya et al. [22, 23] used metal detectors, wet detectors, and dry detectors. For garbage separation, Gundupalli et al. [24] employed a thermal imaging camera, a proximity sensor, and a robotic arm. Huang et al. [25] used an optical sensor to identify the size, location, color, and shape of the garbage as part of an indirect garbage sorting process, through a mechanical separation that is controlled by a computer. A recycling robot in Davidson et al. [26] sorts and classifies waste comprised of plastic, glass, and other materials by using a weight-activated detector and a switch. Pan et al. [27] and Kanta et al. [28] proposed IoT-based systems to detect and classify waste using microcontrollers and multiple sensors.

10.3 Proposed Approach

In this work we propose an approach for garbage detection using deep learning. We explore deep learning for features extraction and visual recognition due to its effectiveness for visual recognition. To minimize the recognition time, we consider the use of standard pre-trained DNN models for feature extraction, due to the availability of hardware implementations of these models. The ConvNet features extracted by the DNN model are used by a classifier for identifying the category of the garbage. The block diagram of the proposed workflow is shown in Figure 10.1.

10.3.1 Pre-Trained DNN for Convnet Feature Extraction

As mentioned in the previous sections, in this work we explore the use of Convolutional Neural Network (CNN) for visual recognition through the use of the extracted ConvNet features. A CNN can be used for a visual classification task in three different ways, as given below.

1. Define the CNN architecture and train it from the scratch
2. Train a classifier on ConvNet features extracted by a pre-trained CNN
3. Train a pre-trained CNN for the new task (transfer learning)

In the above three approaches, transfer learning-based approach and training from scratch need significant training for good results. Compared to the first and third approaches, the second approach of using a pre-trained CNN for ConvNet feature extraction doesn't involve the training of a CNN, thereby only the classifier needs to be trained. The other advantage of the second approach is the availability of hardware implementation of these standard pre-trained models, which reduce the prediction time significantly when compared to software-based computation/extraction of ConvNet features. The next section explains the use of these ConvNet features by a classifier for the recognition task.

10.3.2 Classifier

As explained in the previous section, we used a classifier to identify the garbage category from the images, using the ConvNet features extracted by a pre-trained CNN. To achieve real-time garbage categorization with optimum performance, we have evaluated various linear-classifiers for categorization. The next section explains the experimental study of the proposed approach.

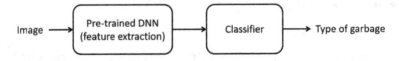

FIGURE 10.1
Block diagram of the proposed workflow.

10.4 Experimental Evaluation

In this section, we discuss the evaluation of the proposed approach on Kaggle's Non and Biodegradable materials dataset. The experimental study was conducted in Matlab R2021a. The next sub-section explains the implementation details of the proposed approach for evaluation on the dataset.

10.4.1 Dataset

This work was evaluated on the recent garbage classification dataset on Kaggle, the Non and Biodegradable materials dataset created by Rayhan Zamzamy [29]. The dataset consists of approximately 156 thousand JPEG images of various image sizes of items which are biodegradable and non-biodegradable in nature. Here, biodegradable items refer to the materials which can be decomposed naturally by microorganisms, such as food and plant products. Non-biodegradable items refer to materials that cannot be decomposed naturally, such as inorganic elements, metals, glass, cement, and plastic. This dataset includes images from Food-101, Waster Classification data v1 and v2, waste pictures, and the Fruit & Vegetable image recognition datasets. The number of images for each category in this dataset for training and testing is shown in Table 10.1.

10.4.2 Results and Analysis

The proposed approach explores the use of pre-trained CNN for ConvNet feature extraction. In this study, we explore the use of AlexNet and ResNet18 for feature extraction. The size of the feature, the time taken for feature extraction, and the performance with linear discriminant analysis for these pre-trained models are shown in Table 10.2.

From the results in Table 10.3, it can be concluded that the best performance of 95.7% is achieved with Quadratic Discriminant Analysis (QDA) classifier. The associated confusion matrix and AUC plots, are shown in Figure 10.2 and Figure 10.3 respectively. The computation of TPR & FNR and PPV & FDR for the proposed model are shown in Figure 10.4 and Figure 10.5

TABLE 10.1

The Number of Observations for Each Category in Train and Test Data of Kaggle's Non- and Biodegradable Materials Dataset

Non- and Biodegradable Materials Dataset	Biodegradable	Non-Biodegradable	Total # of Images
Training Data	119,772	25,569	145,341
Testing Data	2,539	8,259	10,798

TABLE 10.2

The Details of Feature Extraction Using Pre-Trained CNN Models

| Model | ConvNet Feature Size | Feature Extraction Time | | Accuracy using Linear Discriminant Analysis in Percentage |
		Total Time in Sec	Per Image in MSec	
AlexNet	9216	392	2.7	92.8
ResNet18	512	1915	13.18	93.5

TABLE 10.3

Evaluation of Various Classifiers with ResNet18 ConvNet Features

Model	Type	Validation Accuracy	Test Accuracy	Prediction Speed (in Obs/Sec)
Tree	Fine	94.80	87.00	130,000
	Medium	93.70	83.80	130,000
	Coarse	92.00	80.30	130,000
Discriminant Analysis	Linear	97.60	93.50	66,000
	Quadratic	96.90	95.70	67,000

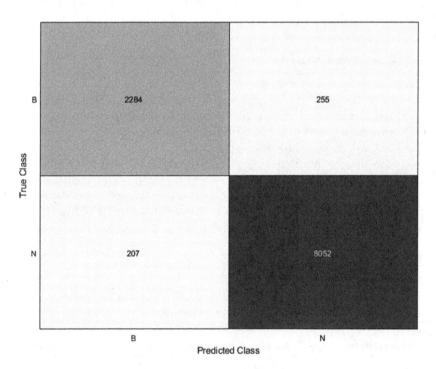

FIGURE 10.2
Confusion matrix of the proposed approach with ResNet18 and QDA Classifier.

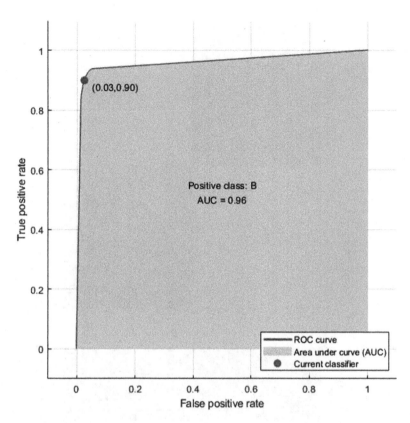

FIGURE 10.3
The ROC and AUC of the proposed approach with ResNet18 and QDA Classifier.

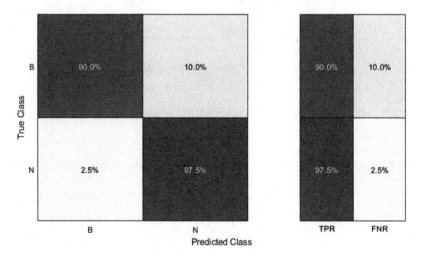

FIGURE 10.4
The True Positive Rate (TPR) and False Negative Rate (FNR) of the proposed approach.

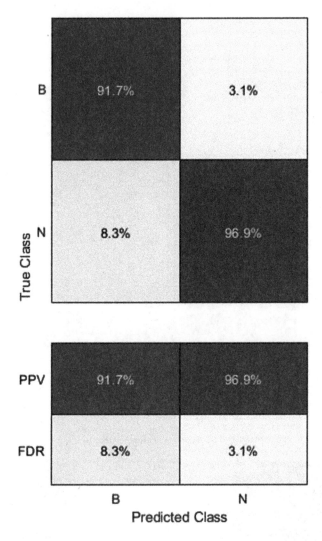

FIGURE 10.5
The Positive Predictive Values (PPV) and False Discovery Rates (FDR) of the proposed approach.

respectively. With a prediction rate of 67,000 images per second, the label of an image is estimated/predicted in 0.0149 msec.

Hence, with ResNet18 feature extraction and QDA classifier as the proposed approach, an image can be classified in approximately 14 milliseconds (13.18+0.0149). This suggests that the proposed approach can classify around 72 images per second. Since the normal frame-rate of video is 24 frames per second, the proposed approach can classify all the frames in the video in real-time.

10.5 Conclusions and Future Work

In this work, we proposed a CNN-based approach to address the problem of garbage classification. The discriminative feature learning capability of CNN and the effective hardware implementation of CNN models are exploited in this work to design an effective real-time garbage classification system using visual data. Various classifiers were analyzed to identify the classifier that achieves optimum results for real-time classification. The experimental study suggests the effectiveness of the proposed approach and confirms the development of a fast garbage classification model. With the hardware implementation of CNN and the low-complexity of the classifier, the proposed approach can be implemented in an IoT environment for real-time operation. The experimental study also suggests the effectiveness of the proposed approach. Future work can explore other DNN architectures and training/optimization strategies to develop a more optimal classification model. The various hardware implementations of the entire workflow can also be explored to create an end-to-end system.

References

1. *Global Waste to Grow by 70 Percent by 2050 Unless Urgent Action is Taken: World Bank Report*, https://www.worldbank.org/en/news/press-release/2018/09/20/global-waste-to-grow-by-70-percent-by-2050-unless-urgent-action-is-taken-world-bank-report.
2. Rad, Mohammad Saeed, Andreas von Kaenel, Andre Droux, Francois Tieche, Nabil Ouerhani, Hazım Kemal Ekenel, and Jean-Philippe Thiran. "A computer vision system to localize and classify wastes on the streets." In Proceedings of the International Conference on Computer Vision Systems, Shenzhen, China, pp. 195–204. Springer, Cham, 2017.
3. Szegedy, Christian, Wei Liu, Yangqing Jia, Pierre Sermanet, Scott Reed, Dragomir Anguelov, Dumitru Erhan, Vincent Vanhoucke, and Andrew Rabinovich. "Going deeper with convolutions." In Proceedings of the IEEE Conference on Computer Vision and Pattern Recognition, Boston, MA, USA, pp. 1–9, 2015.
4. Mittal, Gaurav, Kaushal B. Yagnik, Mohit Garg, and Narayanan C. Krishnan. "Spotgarbage: smartphone app to detect garbage using deep learning." In Proceedings of the ACM International Joint Conference on Pervasive and Ubiquitous Computing, Heidelberg, Germany, pp. 940–945. 2016.
5. Krizhevsky, Alex, Ilya Sutskever, and Geoffrey E. Hinton. "Imagenet classification with deep convolutional neural networks." *Advances in Neural Information Processing Systems*, Lake Tahoe, Nevada, USA, 25 (2012): 1097–1105.
6. Bansal, Siddhant, Seema Patel, Ishita Shah, Prof Patel, Prof Makwana, and D. R. Thakker. "AGDC: Automatic garbage detection and collection." arXiv:1908.05849 (2019), Online: http://arxiv.org/abs/1908.05849.

7. Wang, Ying, and Xu Zhang. "Autonomous garbage detection for intelligent urban management." In Proceedings of the MATEC Web of Conferences, vol. 232, p. 01056. EDP Sciences, 2018.
8. Ren, Shaoqing, Kaiming He, Ross Girshick, and Jian Sun. "Faster r-cnn: Towards real-time object detection with region proposal networks." arXiv:1506.01497 (2015), Online: https://arxiv.org/abs/1506.01497.
9. Zhihong, Chen, Zou Hebin, Wang Yanbo, Liang Binyan, and Liao Yu. "A vision-based robotic grasping system using deep learning for garbage sorting." In Proceedings of the Chinese Control Conference (CCC), Dalian, China, pp. 11223–11226, 2017.
10. Liu, Ying, Zhishan Ge, Guoyun Lv, and Shikai Wang. "Research on automatic garbage detection system based on deep learning and narrowband internet of things." *Journal of Physics: Conference Series*, 1069(1) (2018): 012032. IOP Publishing.
11. Redmon, Joseph, and Ali Farhadi. "YOLO9000: better, faster, stronger." In Proceedings of the IEEE Conference on Computer Vision and Pattern Recognition, Boston, MA, USA, pp. 7263–7271. 2017.
12. Redmon, Joseph, and Ali Farhadi. "Yolov3: An incremental improvement." arXiv:1804.02767 (2018), Online: https://arxiv.org/abs/1804.02767.
13. De Carolis, Berardina, Francesco Ladogana, and Nicola Macchiarulo. "YOLO TrashNet: Garbage detection in video streams." In Proceedings of the IEEE Conference on Evolving and Adaptive Intelligent Systems (EAIS), Bari, Italy, pp. 1–7, 2020.
14. Panwar, Harsh, P. K. Gupta, Mohammad Khubeb Siddiqui, Ruben Morales-Menendez, Prakhar Bhardwaj, Sudhansh Sharma, and Iqbal H. Sarker. "AquaVision: Automating the detection of waste in water bodies using deep transfer learning." *Case Studies in Chemical and Environmental Engineering*, 2 (2020): 100026, pp. 1–5, Sep 2020.
15. Kang, Zhuang, Jie Yang, Guilan Li, and Zeyi Zhang. "An automatic garbage classification system based on deep learning." *IEEE Access*, 8 (2020): 140019–140029.
16. Adedeji, Olugboja, and Zenghui Wang. "Intelligent waste classification system using deep learning convolutional neural network." *Procedia Manufacturing* 35 (2019): 607–612.
17. Yang, Zhihu, and Dan Li. "WasNet: A neural network-based garbage collection management system." *IEEE Access*, 8 (2020): 103984–103993.
18. Rabano, Stephenn L., Melvin K. Cabatuan, Edwin Sybingco, Elmer P. Dadios, and Edwin J. Calilung. "Common garbage classification using mobilenet." In Proceedings of the IEEE International Conference on Humanoid, Nanotechnology, Information Technology, Communication and Control, Environment and Management (HNICEM), Baguio City, Philippines, pp. 1–4, 2018.
19. Vo, Anh H., Minh Thanh Vo, and Tuong Le. "A novel framework for trash classification using deep transfer learning." In *IEEE Access*, 7 (2019): 178631–178639.
20. Aral, Rahmi Arda, Şeref Recep Keskin, Mahmut Kaya, and Murat Hacıömeroğlu. "Classification of trashnet dataset based on deep learning models." In Proceedings of the IEEE International Conference on Big Data (Big Data), Seattle, WA, USA, pp. 2058–2062, 2018.
21. Bircanoğlu, Cenk, Meltem Atay, Fuat Beşer, Özgün Genç, and Merve Ayyüce Kızrak. "RecycleNet: Intelligent waste sorting using deep neural networks." *In* Proceedings of the IEEE Innovations in Intelligent Systems and Applications (INISTA), Thessaloniki, Greece, pp. 1–7, 2018.

22. Chandramohan, Amrutha, Joyal Mendonca, Nikhil Ravi Shankar, Nikhil U. Baheti, Nitin Kumar Krishnan, and M. S. Suma. "Automated waste segregator." In Proceedings of the Texas Instruments India Educators' Conference (TIIEC), Bangalore, India, pp. 1–6, 2014.

23. Sharanya, A., U. Harika, N. Sriya, and Sreeja Kochuvila. "Automatic waste segregator." In Proceedings of the IEEE International Conference on Advances in Computing, Communications and Informatics (ICACCI), Udupi, India, pp. 1313–1319, 2017.

24. Gundupalli Paulraj, Sathish, Subrata Hait, and Atul Thakur. "Automated municipal solid waste sorting for recycling using a mobile manipulator." In Proceedings of the ASME International Design Engineering Technical Conferences and Computers and Information in Engineering Conference, Charlotte, North Carolina, USA, pp 1–10, 2016.

25. Huang, Jiu, Thomas Pretz, and Zhengfu Bian. "Intelligent solid waste processing using optical sensor based sorting technology." In Proceedings of the International Congress on Image and Signal Processing, Yantai, China, pp. 1657–1661, 2010.

26. Davidson, Eric. "The design of an autonomous recycling robot." (2008).

27. Pan, Peng, Junhui Lai, Guorong Chen, Jie Li, Mou Zhou, and Hong Ren. "An intelligent garbage bin based on NB-IOT research mode." In Proceedings of the IEEE International Conference of Safety Produce Informatization (IICSPI), Chongqing, China, pp. 113–117, 2018.

28. Kanta, Sagnik, Srinjoy Jash, and Himadri Nath Saha. "Internet of Things based garbage monitoring system." In Proceedings of the IEEE Annual Industrial Automation and Electromechanical Engineering Conference (IEMECON), Bangkok, Thailand, pp. 127–130, 2017.

29. Non and Biodegradable Materials Dataset, https://www.kaggle.com/rayhan-zamzamy/non-and-biodegradable-waste-dataset

Index